Synthesis Lectures on Mathematics & Statistics

Series Editor

Steven G. Krantz, Department of Mathematics, Washington University, Saint Louis, USA

This series includes titles in applied mathematics and statistics for cross-disciplinary STEM professionals, educators, researchers, and students. The series focuses on new and traditional techniques to develop mathematical knowledge and skills, an understanding of core mathematical reasoning, and the ability to utilize data in specific applications.

David Borthwick

A Primer for Mathematical Analysis

Springer

David Borthwick
Department of Mathematics
Emory University
Atlanta, GA, USA

ISSN 1938-1743　　　　　　ISSN 1938-1751　(electronic)
Synthesis Lectures on Mathematics & Statistics
ISBN 978-3-031-91712-7　　　ISBN 978-3-031-91713-4　(eBook)
https://doi.org/10.1007/978-3-031-91713-4

© The Editor(s) (if applicable) and The Author(s), under exclusive license to Springer Nature Switzerland AG 2025

This work is subject to copyright. All rights are solely and exclusively licensed by the Publisher, whether the whole or part of the material is concerned, specifically the rights of translation, reprinting, reuse of illustrations, recitation, broadcasting, reproduction on microfilms or in any other physical way, and transmission or information storage and retrieval, electronic adaptation, computer software, or by similar or dissimilar methodology now known or hereafter developed.
The use of general descriptive names, registered names, trademarks, service marks, etc. in this publication does not imply, even in the absence of a specific statement, that such names are exempt from the relevant protective laws and regulations and therefore free for general use.
The publisher, the authors and the editors are safe to assume that the advice and information in this book are believed to be true and accurate at the date of publication. Neither the publisher nor the authors or the editors give a warranty, expressed or implied, with respect to the material contained herein or for any errors or omissions that may have been made. The publisher remains neutral with regard to jurisdictional claims in published maps and institutional affiliations.

This Springer imprint is published by the registered company Springer Nature Switzerland AG
The registered company address is: Gewerbestrasse 11, 6330 Cham, Switzerland

If disposing of this product, please recycle the paper.

Preface

Introductory graduate-level analysis courses usually require an undergraduate course in real analysis or advanced calculus as a prerequisite. However, even students who have taken the appropriate prerequisites often feel inadequately prepared. The contents of the prior courses may not line up precisely with the expectations for analysis at a graduate level. Moreover, the material is challenging and lack of retention is a common issue. Finally, even students with a firm grasp on the concepts may not yet be adept at writing analysis proofs. The goal of this book is to help students bridge these gaps.

In my experience of teaching first-year graduate courses on real or complex analysis, the most common issue in student preparation has to do with metric space theory. Most undergraduate analysis courses include an introduction to this topic, and so the key concepts are usually familiar. But that level of exposure often turns out to be insufficient. Metric space topology is the underlying language of mathematical analysis, and to succeed at the graduate level students need to become proficient in this language. This book originated as a set of introductory notes that I wrote for those graduate courses, to help review the core concepts and provide additional practice in analysis proof-writing.

The text covers most of the standard introductory analysis material from the beginning, starting with the completeness of the real numbers. However, the focus is placed on tools and concepts that I expect to need the most review, and the treatment of certain topics is rather streamlined. For example, single-variable calculus is covered only briefly in the final chapter, as a means to show off how certain metric space concepts and tools apply to this context.

The book is intended for self-study, with exercises fully incorporated into the text. These exercises are often used to develop important results which serve as the basis for subsequent proofs and exercises. A complete set of solutions is provided at the end of the book, but I strongly recommended that readers take the time to think through these problems carefully and write out their own proofs before making use of the solutions. Simply reading proofs or looking up solutions contributes little to long-term understanding and retention. Proficiency in the language of analysis is built through active engagement, by puzzling over the concepts and working through strategies for proofs.

Atlanta, USA
February 2025

David Borthwick

Acknowledgments I would like to thank the many analysis students who I have taught in introductory graduate courses over the years. I have benefited greatly from their curiosity and enthusiasm, and their experiences provided the inspiration for this book.

Contents

1 Real Numbers 1
   1.1 The Supremum Property 2
   1.2 Sequences 5
      1.2.1 Monotone Sequences 7
      1.2.2 Upper and Lower Limits 9
   1.3 Cauchy Sequences and Completeness 12

2 Complex Numbers and Series 17
   2.1 Complex Numbers and Sequences 17
   2.2 Series 19
   2.3 Absolute Convergence 21
      2.3.1 Convergence Tests 23
   2.4 Power Series 25

3 Metric Topology 29
   3.1 Metric Spaces 29
   3.2 Open and Closed Sets 33
      3.2.1 Subspace Topology 36
   3.3 Convergence and Completeness 38
      3.3.1 Metric Completion 44
   3.4 Compact Sets 46
      3.4.1 Sequential Compactness 49
   3.5 Baire Category Theorem 52

4 Functions on Metric Spaces 55
   4.1 Continuity 55
      4.1.1 Uniform Continuity 60
   4.2 Sequences of Functions 62
   4.3 Stone-Weierstrass Theorem 65

5	**Real Functions**		71
	5.1 Limits and Continuity		71
		5.1.1 Asymptotics and Order Notation	73
		5.1.2 Continuous Functions	75
	5.2 Differentiation		76
		5.2.1 Differentiation of Power Series	80
		5.2.2 Higher Derivatives	82
	5.3 The Mean Value Theorem		83
		5.3.1 Taylor Approximation	86
		5.3.2 L'Hôpital's Rule	88
	5.4 Integration		90
		5.4.1 Fundamental Theorem of Calculus	94
	5.5 Picard Iteration		95
Solutions to Exercises			99
Bibliography			115
Index			117

Real Numbers 1

The rational numbers are defined as the set of fractions

$$\mathbb{Q} := \left\{ \frac{m}{n} : m, n \in \mathbb{Z}, n \neq 0 \right\}.$$

The standard rules of arithmetic, along with the relations ($<, \leq, >, \leq$) give $\mathbb{Q}$ the structure of an *ordered field*.

To describe quantities that come up in basic geometry, such as π or $\sqrt{2}$, we need to move beyond $\mathbb{Q}$ and allow for irrational numbers. The term *real number* was coined by René Descartes in the 17th century to distinguish the true roots of a polynomial from *imaginary* roots such as $\sqrt{-1}$. The real numbers were assumed to include all rational and irrational values. The development of calculus relied on some intuitive assumptions for the real number system, which could not be formulated precisely because there was no systematic definition of a real number.

This situation was not fully resolved until 1872, when Richard Dedekind and Georg Cantor independently developed constructions of $\mathbb{R}$ as a *completion* of $\mathbb{Q}$. We will discuss both of these constructions in this chapter, without going into the full details. They turn out to lead to equivalent definitions, and in different ways they both reveal a fundamental property that distinguishes the real numbers from the rationals, called *completeness*. It turns out that this notion can be formulated in a great many equivalent ways, and we will encounter a few others in the course of this discussion.

© The Author(s), under exclusive license to Springer Nature Switzerland AG 2025
D. Borthwick, *A Primer for Mathematical Analysis*, Synthesis Lectures on Mathematics & Statistics, https://doi.org/10.1007/978-3-031-91713-4_1

1.1 The Supremum Property

Dedekind constructed real numbers as subsets of $\mathbb{Q}$. An element of $\mathbb{R}$ is a *rational cut*, defined as a subset $A \subset \mathbb{Q}$ such that A and A^c are non-empty, every element of A is strictly less than every element of A^c, and A contains no greatest element. As illustrated Fig. 1.1, the idea is that the right endpoint of the cut represents a number. For $q \in \mathbb{Q}$ the corresponding cut would be $\{r \in \mathbb{Q} : r < q\}$. Irrational numbers are be represented by cuts with no rational endpoint. For example, $\sqrt{2}$ corresponds to $\{r \in \mathbb{Q} : r < 0 \text{ or } r^2 < 2\}$.

To complete Dedekind's construction, we need to define addition, multiplication, and ordering for cuts, and then check that the ordered field axioms are satisfied. For example, addition of cuts is given by adding elements:

$$A_1 + A_2 := \{q_1 + q_2 : q_1 \in A_1, q_2 \in A_2\}.$$

The ordering is also straightforward, defined by inclusion:

$$A_1 \leq A_2 \quad \text{if} \quad A_1 \subset A_2. \tag{1.1}$$

Multiplication of cuts is a bit awkward, because the treatment of positive and negative numbers requires multiple cases. Nevertheless, one can easily make the appropriate definitions and check that the rules of arithmetic are satisfied. We not go into this level of detail here.

One very convenient feature of the cut construction is that the real number system can be extended to include $\pm\infty$ by allowing cuts with one side empty. The empty cut $A = \emptyset$ corresponds to $-\infty$, and the full cut $A = \mathbb{Q}$ represents $+\infty$. This extends $\mathbb{R}$ to define an ordered set called the *extended real numbers* ,

$$\mathbb{R}_\infty := \mathbb{R} \cup \{\pm\infty\}.$$

Since $\emptyset \subset A \subset \mathbb{Q}$ for any cut A, we see that

$$-\infty < x < \infty \tag{1.2}$$

for any $x \in \mathbb{R}$. Note that $\mathbb{R}_\infty$ is merely an ordered set and not a field. That is, the arithmetic of $\mathbb{R}$ does not extend because expressions such as $\infty - \infty$ or $0 \cdot \infty$ cannot be consistent defined.

Dedekind's approach to the real numbers naturally leads to a completeness property expressed in terms of the ordering. An extended real number $\beta \in \mathbb{R}_\infty$ is called an *upper*

Fig. 1.1 A rational cut $A \subset \mathbb{Q}$ representing a real number x

1.1 The Supremum Property

bound for a set $E \subset \mathbb{R}$ if $x \leq \beta$ for all $x \in E$. The set E has a *supremum* (or least upper bound) $\alpha \in \mathbb{R}_\infty$ if $\alpha \in \mathbb{R}_\infty$ is an upper bound for E such that every other upper bound β satisfies $\beta \geq \alpha$. If it exists, the supremum of E is clearly unique and is written

$$\alpha = \sup E.$$

By (1.2), $\sup E = \infty$ means that E is not bounded above by a real number, and $\sup E = -\infty$ if and only if E is empty.

The supremum of a set is easily obtained by taking the union of the corresponding cuts. If a point $x \in \mathbb{R}$ is represented by the cut $A_x \subset \mathbb{Q}$, then for $E \subset \mathbb{R}$ we set

$$\sup E := \bigcup_{x \in E} A_x. \qquad (1.3)$$

One must check that the union on the right qualifies as a cut, but this is straightforward. Because the ordering is defined by inclusion, one can easily verify that (1.3) has the properties that characterize $\sup E$. This definition applies to any subset E, which yields the following fundamental result:

Theorem 1.1 (supremum property) *Every subset of $\mathbb{R}$ has a supremum in $\mathbb{R}_\infty$.*

Since we are not going into the full details of the construction, we will accept the result of Theorem 1.1 an axiom. An ordered field with the supremum property is called *complete*, and it turns out that $\mathbb{R}$ is uniquely characterized by this description. That is, any ordered field extension of $\mathbb{Q}$ with the supremum property is isomorphic to $\mathbb{R}$.

The reverse of the supremum is the *infimum*, or (greatest lower bound). For a set $E \subset \mathbb{R}$, the value $\alpha = \inf E$ exists if E has a *lower bound* $\alpha \in \mathbb{R}_\infty$ such that every lower bound β satisfies $\beta \leq \alpha$. Since the reflection $x \mapsto -x$ interchanges upper bounds with lower bounds, Theorem 1.1 implies also that every subset of $\mathbb{R}$ has an infimum in $\mathbb{R}_\infty$. The statement $\inf E = -\infty$ means E not bounded below by a real number, and $\inf E = \infty$ if and only if E is empty.

If $\sup E \in E$ then we say that E has a *maximum* element and use the notation $\max E$ in place of $\sup E$ to signify this. Similarly, if $\inf E \in E$ then E has a *minimum* element denoted by $\min E$.

The *Archimedean property* of $\mathbb{R}$ is the statement that the set of integers $\mathbb{Z}$ does not have a real upper bound. Because we are used to picturing the real numbers as a line in which the integers are embedded, this might seem to be an obvious fact. However, the Archimedean property does not follow from the ordered field axioms alone. (For example, the ordered field of rational functions over $\mathbb{R}$ is not Archimedean.) The point we want to make here is that the Archimedean property follows from the supremum property, so that Dedekind's definition of $\mathbb{R}$ agrees with our intuitive picture of the number line.

Theorem 1.2 (Archimedean property) *Given $x \in \mathbb{R}$, there exists $n \in \mathbb{Z}$ such that $n > x$.*

Proof Assume, for the sake of contradiction, that $\sup \mathbb{Z} = a < \infty$. By the definition of the supremum, $a - 1$ is not an upper bound for $\mathbb{Z}$, and hence there exists some $k \in \mathbb{Z}$ such that $k > a - 1$. But then $k + 1 > a$, which contradicts the fact that a is an upper bound for the integers. $\square$

As a corollary, Theorem 1.2 implies that for any real $\varepsilon > 0$ there exists $n \in \mathbb{N}$ such that
$$\frac{1}{n} < \varepsilon$$
(since $1/\varepsilon$ would be an upper bound for $\mathbb{Z}$ if this were false). We can extend this result to derive the following:

Theorem 1.3 *Given $x, y \in \mathbb{R}$ with $x < y$, there exists $q \in \mathbb{Q}$ such that $x < q < y$.*

Proof For $x < y$ the Archimedean property allows us to choose $n \in \mathbb{N}$ so that
$$\frac{1}{n} < y - x.$$
Let
$$m = \inf\{k \in \mathbb{Z} : k > nx\},$$
and observe that the set $\{k > nx\}$ is not empty, by the Archimedean principle, but is bounded below. Therefore m is a finite integer. Since $nx + 1 < ny$, m satisfies
$$nx < m < ny.$$
Setting $q = m/n$ yields the result. $\square$

We conclude this section with a review of the basic terminology of intervals. An *interval* I is defined as a convex subset of $\mathbb{R}$, which means that if $x, y \in I$ then $t \in I$ for every point $x \leq t \leq y$. The empty set and the single point $\{x\}$ qualify as intervals by default. Given a non-empty interval I we can associate endpoints
$$\alpha = \inf I, \quad \beta = \sup I.$$
It follows from the definitions that $x \in I$ for all $\alpha < x < \beta$. Real endpoints may or may not be included in I, and infinite endpoints are automatically excluded. The interval is *bounded* if both endpoints are real.

An interval is said to be *open* if it is either empty or does not include either endpoint. The exclusion is indicated by parentheses,

$$(\alpha, \beta) := \{x \in \mathbb{R} : \alpha < x < \beta\}.$$

An interval is *closed* if it is either empty or includes its real endpoints. Square brackets are used to indicate this,

$$[a, b] := \{x \in \mathbb{R} : a \leq x \leq b\}.$$

The infinite intervals $(-\infty, b]$, $[a, \infty)$, and $(-\infty, \infty)$ are also closed by this definition. Note that $\emptyset$ and $\mathbb{R}$ are classified as both open and closed.

1.2 Sequences

A sequence in $\mathbb{R}$ is an ordered list of numbers $(x_1, x_2, x_3, \dots)$. We will write this in compact form as $(x_n)_{n \in \mathbb{N}}$, or simply (x_n). We say that (x_n) *converges* to a limit $y \in \mathbb{R}$ if for every $\varepsilon > 0$,

$$|x_n - y| < \varepsilon \quad \text{for all but finitely many } n. \tag{1.4}$$

This is notated as

$$\lim_{n \to \infty} x_n = y,$$

or simply $x_n \to y$ when there is no ambiguity in the index. The statement (1.4) can be made more explicit by saying that for each $\varepsilon > 0$ there exists N so that $|x_n - y| < \varepsilon$ for all $n \geq N$.

Example 1.4 The sequence $n^{1/n}$ plotted in Fig. 1.2 appears to be converging to 1. To verify this, set $x_n = n^{1/n} - 1$ and note that $x_n \geq 0$ and

$$(1 + x_n)^n = n. \tag{1.5}$$

Extracting the quadratic term from the binomial expansion of $(1 + x_n)^n$ then gives an estimate

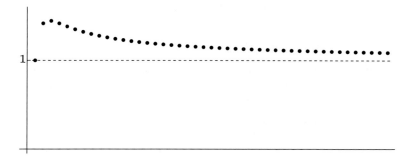

Fig. 1.2 A plot of the sequence $n^{1/n}$

$$\frac{n(n-1)}{2}x_n^2 \leq n.$$

This yields the bound

$$\left|n^{1/n} - 1\right| \leq \sqrt{\frac{2}{n-1}}. \tag{1.6}$$

Given $\varepsilon > 0$, if we choose $N > 1 + 2/\varepsilon^2$, then $n \geq N$ implies $|x_n| < \varepsilon$. $\Diamond$

It makes no difference to the definition (1.4) if the inequality $|x_n - y| \leq \varepsilon$ is used in place of the strict inequality, because the statement needs to hold for all $\varepsilon > 0$. By the same reasoning, we can replace ε with some other expression, such as 2ε, as long as it is immediately clear that the estimate can be made arbitrarily small through our choice of N.

Because the definition of the limit is a conditional statement, it is worth considering explicitly what it means for a limit to fail to exist. The condition "for every $\varepsilon > 0$" fails if there is at least one exceptional case. Thus, $x_n \not\to y$ means that there is at least one value of $\varepsilon > 0$ such that $|x_n - y| < \varepsilon$ for only finitely many n.

Some basic but essential properties of sequence limits are highlighted in the following:

Exercise 1.5 Show that a sequence (x_n) that converges in $\mathbb{R}$ is *bounded*, meaning there exists M so that $|x_n| \leq M$ for all n.

Exercise 1.6 If $x_n \to a$ and $y_n \to b$ in $\mathbb{R}$, prove that

$$\lim_{n \to \infty} (x_n + y_n) = a + b \quad \text{and} \quad \lim_{n \to \infty} (x_n y_n) = ab.$$

So far we have been talking only about finite limits. Sequences that do not converge in $\mathbb{R}$ may still have a meaningful limit in the extended real numbers $\mathbb{R}_\infty$. We write

$$\lim_{n \to \infty} x_n = \infty,$$

or $x_n \to \infty$, if for each $m \in \mathbb{R}$ all but finitely many points of the sequence satisfy $x_n \geq m$. Similarly, we write

$$\lim_{n \to \infty} x_n = -\infty,$$

or $x_n \to -\infty$, if for each $m \in \mathbb{R}$ all but finitely many points satisfy $x_n \leq m$. These extended definitions are quite useful, but we do need to be careful with our use of the term *convergent*. Normally, a sequence is called convergent only if the limit is finite. However, this usage may vary depending on the context. When there is a possibility for confusion, we will describe the limits as being "in $\mathbb{R}$" or "in $\mathbb{R}_\infty$."

Exercise 1.7 Let A be a non-empty subset of $\mathbb{R}$. Prove that there exists a sequence (x_n) in A with $x_n \to \sup A$. (The same results holds for the infimum.)

1.2 Sequences

The algebraic properties of Exercise 1.6 do not apply to extended limits, because arithmetic is restricted to $\mathbb{R}$. We can apply ordering relations to extended limits, however.

Exercise 1.8 Suppose that (x_n) and (y_n) are sequences in $\mathbb{R}$ with limits in $\mathbb{R}_\infty$, and that $x_n \leq y_n$ for all n. Prove that
$$\lim x_n \leq \lim y_n.$$

Note that the results of Exercises 1.6 and 1.8 require that the limits are already known to exist. This creates a potential pitfall in proofs where the goal is to prove existence of a limit. We need to be careful not to assume this existence prematurely.

Example 1.9 Consider the Fibonacci sequence $0, 1, 1, 2, 3, 5, \ldots$, which is defined recursively by setting $x_0 = 0$, $x_1 = 1$, and
$$x_n = x_{n-1} + x_{n-2}.$$

It should be clear that $x_n \to \infty$. However, if we simply apply the rules from Exercise 1.6, without checking existence, we would deduce that
$$\lim x_n = 2 \lim x_n.$$

This leads to the nonsensical conclusion that $\lim x_n = 0$. ◊

1.2.1 Monotone Sequences

Real limits are relatively easy to handle if we can avoid oscillations. The term for a sequence without oscillation is *monotone*. A monotone sequence (x_n) is either *increasing* ($x_n \leq x_{n+1}$) or *decreasing* ($x_n \geq x_{n+1}$).

Theorem 1.10 *A monotone sequence in $\mathbb{R}$ has a limit in $\mathbb{R}_\infty$.*

Proof It suffices to consider the increasing case, since the same argument can be applied to a decreasing sequence by replacing x_n by $-x_n$. For an increasing sequence (x_n), we claim that the limit is given by
$$\alpha := \sup\{x_1, x_2, \ldots\}.$$
An example is illustrated in Fig. 1.3.

To see that $x_n \to \alpha$, suppose first that $\alpha \in \mathbb{R}$. For $\varepsilon > 0$, $\alpha - \varepsilon$ is not a lower bound on $\{x_n\}$. Therefore $x_N \geq \alpha - \varepsilon$ for some N. Since the sequence is increasing and α is an upper bound, we have
$$x_n \in [\alpha - \varepsilon, \alpha] \quad \text{for all } n \geq N.$$

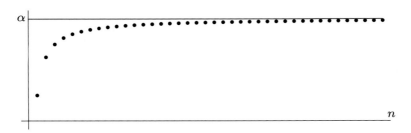

Fig. 1.3 An increasing sequence (x_n) converging to $\alpha = \sup x_n$

This shows that $x_n \to \alpha$ in the finite case.

Since $\{x_n\}$ is not empty, $\alpha \neq -\infty$ and the other possibility to consider is $\alpha = \infty$. In this case, for any $m \in \mathbb{R}$ we know that m is not an upper bound for $\{x_n\}$. This implies that $x_N \geq m$ for some N, and hence $x_n \geq m$ for all $n \geq N$ since the sequence is increasing. □

Example 1.11 Consider a decimal expansion $0.d_1d_2d_3,\ldots$, with digits $d_j \in \{0, 1, \ldots, 9\}$. This corresponds to a sequence of rational numbers given by

$$x_n = \frac{d_1}{10} + \frac{d_2}{100} + \cdots + \frac{d_n}{10^n}.$$

The sequence (x_n) is increasing because the terms in the sum are positive. Therefore $x = \lim x_n$ exists and is contained in $[0, 1]$ since $0 \leq x_n \leq 1$ for all n. In other words, each decimal expansion represents a real number. ◊

Example 1.12 Newton's approximation method gives a rational approximation to $\sqrt{2}$ defined by the sequence $2, \frac{3}{2}, \frac{17}{12}, \ldots$. This is defined iteratively by setting $x_1 = 2$ and

$$x_{n+1} = \frac{x_n^2 + 2}{2x_n}. \tag{1.7}$$

Using induction, we can deduce from (1.7) that

$$\sqrt{2} \leq x_n \leq 2$$

for all n, and also that (x_n) is a decreasing sequence. Theorem 1.10 implies that the limit exists in $\mathbb{R}$. Taking the limit on both sides of (1.7) shows that $\lim x_n = \sqrt{2}$. ◊

The monotone sequence theorem is actually another version of the completeness property for $\mathbb{R}$, equivalent to the supremum property. We can use it to deduce the following result for nested sequences of intervals, which is yet another form of completeness. We will see this generalized to metric spaces later on in Sect. 3.4.

1.2 Sequences

Lemma 1.13 *Suppose I_n is bounded, closed, nonempty interval for each n, and that these intervals form a nested sequence,*

$$I_1 \subset I_2 \subset I_3 \subset \ldots.$$

Then the intersection $\cap I_n$ is not empty.

Proof If we write $I_n = [a_n, b_n]$, then the nesting property means that (a_n) is increasing and (b_n) is decreasing. These sequences are bounded by $[a_1, b_1]$, so they have real limits

$$a = \lim a_n, \quad b = \lim b_n.$$

Since $a_m \leq b_n$ for all m and n, it follows from Exercise 1.8 that $a \leq b$ and $[a, b] \in I_n$ for all n. □

As an example of how Lemma 1.13 can be used, let us present a proof of the uncountability of the real numbers. Recall that an infinite set E is *countable* if there exists a bijection $E \leftrightarrow \mathbb{N}$. This is equivalent to saying that the elements can be listed as a sequence,

$$E = \{x_1, x_2, \ldots\}.$$

An infinite set that cannot be listed as a sequence is called *uncountable*.

Theorem 1.14 *The set $\mathbb{R}$ is uncountable.*

Proof Suppose, for the sake of contradiction, that $\mathbb{R}$ is countable, so that we can write

$$\mathbb{R} = \{x_1, x_2, \ldots\}. \tag{1.8}$$

Choose some closed interval I_1 such that $x_1 \notin I_1$. Then find a closed interval $I_2 \subset I_1$ such that $x_2 \notin I_2$, and so on. Continuing this process yields a nested sequence

$$I_1 \supset I_2 \supset I_3 \supset \ldots.$$

The intersection $\cap I_n$ is not empty by Lemma 1.13. However, since $x_k \notin I_k$ for each k, this contradicts (1.8). □

1.2.2 Upper and Lower Limits

A typical strategy for proving that $x_n \to y$ involves estimating the difference $|x_n - y|$. Such estimates typically require algebraic operations and comparisons. Thus the potential pitfall, as noted in the discussion of Exercises 1.6 and 1.8, is that we can only apply algebraic operations and ordering relations to the limits if we already know that the limit exists.

To work around this we use upper and lower limits, whose existence is guaranteed. Given a real sequence (x_n), for each $n \in \mathbb{N}$ define

$$y_n = \sup\{x_k : k \geq n\}.$$

Then y_n also serves an upper bound on $\sup\{x_k : k \geq n+1\}$. This implies $y_n \geq y_{n+1}$, so the sequence is decreasing. By Theorem 1.10, we can take the limit in $\mathbb{R}_\infty$ to define

$$\limsup x_n := \lim_{n\to\infty} y_n.$$

The definition could be written more succinctly as

$$\limsup x_n := \lim_{n\to\infty} \left(\sup_{k\geq n} x_k\right),$$

(which explains the terminology). Similarly, we define

$$\liminf x_n := \lim_{n\to\infty} \left(\inf_{k\geq n} x_k\right).$$

Note that the infimum of a set is always less than the supremum, and so

$$\liminf x_n \leq \limsup x_n \tag{1.9}$$

by Exercise 1.8.

Example 1.15 Consider the alternating sequence

$$x_n = (-1)^n \frac{n+1}{n},$$

illustrated in Fig. 1.4. For all n we can estimate

$$1 \leq \sup_{k\geq n} x_k \leq 1 + \frac{1}{n},$$

which gives $\limsup x_n = 1$. Similarly, $\liminf x_n = -1$. ◊

The lim sup and lim inf do not provide direct bounds on elements of the sequence. For example, in the sequence plotted in Fig. 1.4, infinitely points lie above the lim sup and below the lim inf, with none in between. In order to produce inequalities related to the upper and lower limits, we must move slightly above or below these values. The following characterization makes this notion explicit:

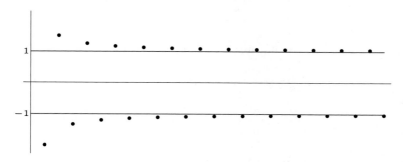

Fig. 1.4 The upper and lower limits of the sequence from Example 1.15

Exercise 1.16 For a real sequence (x_n), show that $\limsup x_n$ is equal to the unique value $\alpha \in \mathbb{R}_\infty$ satisfying, for all $c \in \mathbb{R}$:

(i) If $c > \alpha$ then $x_n > c$ for only finitely many n.
(ii) If $c < \alpha$, then $x_n > c$ for infinitely many n.

(These conditions are reversed for $\liminf x_n$.)

This formulation is the key to many applications of lim sup and lim inf. For example, we can easily derive the following fundamental result:

Lemma 1.17 *A real sequence* (x_n) *has a limit* $\alpha \in \mathbb{R}_\infty$ *if and only if*

$$\alpha = \liminf x_n = \limsup x_n$$

Proof If both values $a = \liminf x_n$ and $b = \limsup x_n$ are real, then we an paraphrase the result of Exercise 1.16 in terms of an arbitrary $\varepsilon > 0$:

(i') The interval $[a - \varepsilon, b + \varepsilon]$ contains all but finitely many x_n.
(ii") Infinitely many x_n lie outside $(a + \varepsilon, b - \varepsilon)$.

If $a = b$ then (i') is equivalent to the statement that $x_n \to a$, while (ii') is vacuous. And if $a < b$ then (ii') shows that no limit exists.

The infinite cases follow directly from Exercise 1.16. For example, if $\limsup x_n = -\infty$ then (i) says that for all $c \in \mathbb{R}$, $x_n > c$ for only finitely many n. This is the same as the definition of $x_n \to -\infty$. □

Upper and lower limits are particularly useful for estimates because they exist for any sequence and satisfy ordering relations.

Exercise 1.18 If (x_n) and (y_n) are sequences in $\mathbb{R}$ with $x_n \leq y_n$ for all n, show that

$$\liminf x_n \leq \liminf y_n, \quad \limsup x_n \leq \limsup y_n. \tag{1.10}$$

We need to be a bit more careful with algebraic relations for lim sup and lim inf, because the results are one-sided.

Exercise 1.19 For bounded real sequences (x_n) and (y_n), prove that

$$\limsup (x_n + y_n) \leq \limsup x_n + \limsup y_n. \tag{1.11}$$

(The reverse inequality holds for lim inf.)

It is not difficult to see that equality will hold in (1.11) when at least one of the sequences has a limit. Strict inequality is certainly possible, though. For example, if $x_n = (-1)^n$ and $y_n = (-1)^{n+1}$, then $\limsup x_n = \limsup y_n = 1$, whereas $x_n + y_n = 0$ for all n,

The lim sup of a sequence of products may not be related to the lim sup of the factors in general. However, when one of the factor sequences converges to a finite positive limit, then the limit can be extracted:

Exercise 1.20 Let (x_n) and (x_n) be bounded positive sequences and suppose that $x_n \to a$. Prove that

$$\limsup (x_n y_n) = a \limsup y_n.$$

(The same result holds for lim inf.)

A *subsequence* of (x_n) is a new sequence of the form $(x_{n_k})_{k \in \mathbb{N}}$, where $n_k \in \mathbb{N}$ with $n_k < n_{k+1}$. If the sequence has a limit, then it is clear that any subsequence will approach the same limit. Furthermore, whether or not the full sequence has a limit, we can always find subsequences that do.

Exercise 1.21 Prove that a real sequence (x_k) has a subsequence such that

$$\lim_{n \to \infty} x_{k_n} = \limsup x_k.$$

(The same result holds for $\liminf x_k$, of course.)

1.3 Cauchy Sequences and Completeness

The definition of completeness in terms of the supremum property applies to any ordered field, but as we remarked after Theorem 1.1, the real number system is the only ordered field with this property. This definition of completeness thus has an extremely limited scope.

1.3 Cauchy Sequences and Completeness

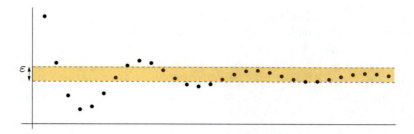

Fig. 1.5 A sequence is Cauchy if all but finitely points are contained in an interval of arbitrarily small width ε

We have already remarked on some other notions of completeness that yield an equivalent axiom for the real numbers. In this section we will discuss yet another, the concept of *metric completeness*. This property was first adopted as an axiom by Augustin-Louis Cauchy in the early 19th century, as part of his work on rigorous foundations of calculus.

A sequence (x_n) in $\mathbb{R}$ is said to be *Cauchy* if for every $\varepsilon > 0$, all but finitely many points of the sequence lie within an interval of width ε. An equivalent way to say this is that for each $\varepsilon > 0$ there exists N so that

$$|x_n - x_m| < \varepsilon \quad \text{for all } n, m \geq N.$$

This Cauchy condition is illustrated in Fig. 1.5. This sounds similar to the definition of a limit, but the crucial point here is that no limiting value is specified. If $x_n \to y$, then we could pick $(y - \frac{\varepsilon}{2}, y + \frac{\varepsilon}{2})$ for the interval required by the Cauchy condition. This observation yields the following:

Lemma 1.22 *A sequence that converges in $\mathbb{R}$ is Cauchy.*

A space is *complete* in the metric sense if all Cauchy sequences converge. We will demonstrate that $\mathbb{R}$ has this property as a consequence of the existence of upper and lower limits.

Theorem 1.23 *All Cauchy sequences in $\mathbb{R}$ are convergent.*

Proof Suppose (x_n) is a Cauchy sequence in $\mathbb{R}$. Given $\varepsilon > 0$, the fact that there is an interval of width ε that contains all but finitely many x_n shows that the sequence is bounded, with

$$\limsup x_n - \liminf x_n \leq \varepsilon.$$

Since ε was arbitrary, we conclude that $\limsup x_n = \liminf x_n$, and convergence follows from Lemma 1.17. □

It is easy to see that $\mathbb{Q}$ is not complete in the metric sense. For example, the decimal expansions discussed in Example 1.11 clearly correspond to Cauchy sequences in $\mathbb{Q}$, because decimal expansions which agree out to the nth decimal place yield numbers that differ by at most 10^{1-n}. These sequences do not converge $\mathbb{Q}$ if the decimal expansion is non-repeating.

Cantor's construction of real numbers is explicitly based on Cauchy sequences. A real number is defined as an equivalence class of Cauchy sequences in $\mathbb{Q}$, under the equivalence relation

$$(x_n) \sim (y_n) \text{ if } \lim_{n \to \infty} (x_n - y_n) = 0.$$

This approach can be generalized to the concept of *metric space completion*, which we will discuss later in Sect. 3.3.1.

The main distinction between this construction and Dedekind's is that metric completeness is essentially built into Cantor's approach, just as the supremum property is an immediate consequence of Dedekind's. Although we have not gone into the details of either construction, we will at least show that the axioms are equivalent.

Theorem 1.24 *For $\mathbb{R}$ the supremum property is equivalent to metric completeness.*

Proof The forward implication was already shown in Theorem 1.23, so we assume that $\mathbb{R}$ is an ordered field in which Cauchy sequences in are convergent, but for which the supremum property is not yet established. For $E \subset \mathbb{R}$ our goal is to prove that $\sup E$ exists in $\mathbb{R}_\infty$. The infinite cases are straightforward. If E is empty then $\sup E = -\infty$ and if E has no real upper bound then $\sup E = \infty$. This leaves the case where E is not empty and has an upper bound in $\mathbb{R}$.

Under these assumptions on E, we claim that for each $\varepsilon > 0$ there exists a point $x \in E$ such that $x + \varepsilon$ is an upper bound for E. To see this, let m be an upper bound on E, and pick a starting point $x_1 \in E$. If $x_1 + \varepsilon$ is an upper bound then we are done. Otherwise, there exists some point $x_2 \in E$ with $x_2 > x_1 + \varepsilon$. We can continue this process until we find $x_k \in E$ such that $x_k + \varepsilon$ is an upper bound. This will happen in at most $k \leq (m - x_1)/\varepsilon$ steps.

With the claim proven, we can now show that a set E that is not empty and bounded above has a supremum. For each n, use the claim above to find $x_n \in E$ such that $x_n + 1/n$ is an upper bound for E. For any two points x_n and x_m in the sequence,

$$x_n \leq x_m + \frac{1}{m} \quad \text{and} \quad x_m \leq x_n + \frac{1}{n}.$$

For all $m, n \geq N$ this implies that

$$|x_n - x_m| \leq \frac{1}{N},$$

and thus the sequence is Cauchy. By hypothesis (x_n) converges in $\mathbb{R}$.

The final step is to show that $a := \lim x_n$ is the supremum of E. If $x \in E$, then

1.3 Cauchy Sequences and Completeness

$$x \leq x_n + \frac{1}{n}$$

for all n. Since $x_n \to a$ and $1/n \to 0$ as $n \to \infty$, this implies that $x \leq a$. Therefore a is an upper bound for E. If $\varepsilon > 0$, then $x_n > a - \varepsilon$ for all but finitely many n, and so $a - \varepsilon$ is not an upper bound for E. This proves that sup E exists and is equal to a. □

We can now verify the comment made earlier about the monotone sequence theorem. The convergence of monotone sequences implies the existence of the lim sup and lim inf, and we used that fact to establish metric completeness. Theorem 1.24 shows that the supremum property follows from this.

To conclude this section, let us consider one final version of the completeness axiom, related to subsequences.

Theorem 1.25 (Bolzano-Weierstrass) *A bounded sequence in $\mathbb{R}$ has a convergent subsequence.*

This follows immediately from Exercise 1.21, since a bounded sequence has a finite lim sup. We leave it to the reader to check the equivalence to metric completeness.

Exercise 1.26 Prove the converse of Theorem 3.26, i.e., the Bolzano-Weierstrass property implies that all Cauchy sequences are convergent.

Complex Numbers and Series

Algebraic operations such as sums involve only a finite number of terms by definition. The term *series* refers to the limiting case where the number of summands is taken to infinity. The goal of this chapter is to develop some of the basic theory of numerical series. We start with a brief introduction to the field of complex numbers, since that is the natural context for many of the fundamental results.

2.1 Complex Numbers and Sequences

The field of complex numbers is defined by *adjoining* to $\mathbb{R}$ an element i which satisfies $i^2 = -1$. In terms of the additive structure this means that $\mathbb{C}$ is a vector space over $\mathbb{R}$, spanned by 1 and i:
$$\mathbb{C} := \{x + iy : x, y \in \mathbb{R}\}.$$

Multiplication is extended to $\mathbb{C}$ by further assuming that i commutes with all other elements and satisfies the distributive law. This gives the multiplication rule:
$$(a + ib)(c + id) := (ac - bd) + i(ad + bc).$$

With these definitions, $\mathbb{C}$ is an algebra containing $\mathbb{R}$ as a subalgebra, with additive unit 0 and multiplicative unit 1. To establish that $\mathbb{C}$ is a field, it remains to check the existence of multiplicative inverses, which we will take care of below.

The real and imaginary parts of a complex number $z = x + iy$ are notated as
$$\operatorname{Re} z := x, \quad \operatorname{Im} z := y$$

and the *conjugate* is defined as

© The Author(s), under exclusive license to Springer Nature Switzerland AG 2025
D. Borthwick, *A Primer for Mathematical Analysis*, Synthesis Lectures
on Mathematics & Statistics, https://doi.org/10.1007/978-3-031-91713-4_2

$$\bar{z} := x - iy.$$

The absolute value is extended from $\mathbb{R}$ to $\mathbb{C}$ by defining

$$|z| := \sqrt{z\bar{z}} = \sqrt{x^2 + y^2}, \qquad (2.1)$$

Note that this agrees with the Euclidean norm on $\mathbb{R}^2$. It follows that the complex absolute value is positive definite and satisfies the triangle inequality,

$$|z + w| \leq |z| + |w|$$

for all $z, w \in \mathbb{C}$. Moreover, (2.1) implies that the absolute value is multiplicative,

$$|zw| = |z||w|,$$

just as in the real case.

To complete the description of $\mathbb{C}$ as a field, the fact that $|z|^2 = z\bar{z}$ gives a simple formula for the multiplicative inverse,

$$z^{-1} = \frac{\bar{z}}{|z|^2}$$

for $z \neq 0$. It is straightforward to verify that the field axioms are satisfied, and we will not go into those details here.

The complex number field cannot be ordered, because $i^2 = -1$ contradicts the ordered field axioms. Hence, only real numbers can appear in inequalities. This an assumption that is often made implicitly. For example, use of the phrase "given $\varepsilon > 0$" implies that $\varepsilon \in \mathbb{R}$.

Sequential limits in $\mathbb{C}$ are defined by using the absolute value to measure distance between points, just as for real numbers. That is, $z_k \to z$ means that for every $\varepsilon > 0$, there exists N so that

$$|z_k - z| < \varepsilon \quad \text{for all } k \geq N.$$

This is related to real convergence by the following:

Exercise 2.1 A complex sequence converges if and only if its sequences of real and imaginary parts converge separately in $\mathbb{R}$. ◊

By Exercise 2.1, the algebraic properties of sequential limits noted for $\mathbb{R}$ in Exercise 1.6 carry over directly to $\mathbb{C}$.

Lemma 2.2 *Suppose (w_k) and (z_k) are convergent sequences in $\mathbb{C}$ with $w_k \to w$ and $z_k \to z$. Then*

$$\lim (w_k + z_k) = w + z \quad \text{and} \quad \lim w_k z_k = wz.$$

A complex sequence (z_n) is *Cauchy* if for every $\varepsilon > 0$, there exists N such that

2.2 Series

$$|z_n - z_m| < \varepsilon \quad \text{for all } n, m \geq N.$$

As in the real case, the triangle inequality implies that convergent sequences are Cauchy. The converse also holds, by the following extension of Theorem 1.23.

Exercise 2.3 Prove that all Cauchy sequences in $\mathbb{C}$ are convergent.

2.2 Series

Formally, a complex series is an infinite sum

$$z_0 + z_1 + z_2 + \ldots,$$

with each $z_k \in \mathbb{C}$. The series *converges* if the sequence of partial sums given by

$$s_n := z_0 + \cdots + z_n$$

has a limit in $\mathbb{C}$. In this case we define

$$\sum_{k=0}^{\infty} z_k := \lim_{n \to \infty} s_n.$$

The series is said to *diverge* if the sequence of partial sums has no limit. For a real series, we can defined extended limits $\sum x_n = \pm\infty$, but these infinite cases are still classified as divergent series.

Because a series $\sum z_k$ has two associated sequences, namely the sequence of partial sums (s_n) and the sequence of terms (z_k), when discussing convergence it is important to keep these straight.

Exercise 2.4 Prove that if $\sum z_k$ converges, then $\lim z_k = 0$.

The converse of the result of Exercise 2.4 is false, as demonstrated by the following:

Example 2.5 Consider the *harmonic* series

$$1 + \frac{1}{2} + \frac{1}{3} + \cdots.$$

We can group the terms by powers of two to estimate the partial sums:

$$s_{2^m} = 1 + \frac{1}{2} + \left(\frac{1}{3} + \frac{1}{4}\right) + \cdots + \left(\frac{1}{2^{m-1}+1} + \cdots + \frac{1}{2^m}\right)$$
$$\leq 1 + \underbrace{\frac{1}{2} + \cdots + \frac{1}{2}}_{m}$$

This implies that $s_{2^m} \to \infty$, and therefore $s_n \to \infty$, since the sequence of partial sums is increasing. The harmonic series is thus divergent, with

$$\sum_{k=1}^{\infty} \frac{1}{k} = \infty.$$

◇

One of the most useful tools for the analysis of series is a particular case that can be evaluated in closed form:

Example 2.6 For the *geometric* series

$$1 + z + z^2 + \ldots,$$

divergence is clear if $|z| \geq 1$, by Exercise 2.6. To establish convergence for $|z| < 1$, the partial sums can be computed using the polynomial identity

$$(1-z)(1 + z + \cdots + z^n) = 1 - z^{n+1}.$$

For $z \neq 1$ this gives

$$\sum_{k=0}^{n} z^k = \frac{1 - z^{n+1}}{1 - z}.$$

Since $|z|^n \to 0$ as $n \to \infty$ for $|z| < 1$, the series converges to

$$\sum_{k=0}^{\infty} z^k = \frac{1}{1-z}. \qquad (2.2)$$

◇

We can immediately derive some basic algebraic properties of series from the properties established for sequences in Lemma 2.2.

Lemma 2.7 *If $\sum z_k$ and $\sum w_k$ are convergent series, then*

$$\sum (z_k + w_k) = \sum z_k + \sum w_k,$$

and, for $c \in \mathbb{C}$,
$$\sum cz_k = c \sum z_k.$$

Note that multiplication of series is not included in Lemma 2.7. The distribution of terms in a product of two series results in a double summation that is not well-defined without some extra assumptions. (See Exercise 2.13 below.)

By the completeness property of Exercise 2.3, a series converges in $\mathbb{C}$ if and only if the sequence of partial sums is Cauchy. This gives the following criterion:

Theorem 2.8 *A complex series $\sum z_k$ converges if and only if for each $\varepsilon > 0$ there exists $N \in \mathbb{N}$ such that*
$$|z_n + \cdots + z_m| \leq \varepsilon$$
for all $m \geq n \geq N$.

The monotone sequence theorem implies another series convergence result. For a series with real positive terms, the sequence of partial sums is increasing. Hence, Theorem 1.10 yields the following corollary:

Theorem 2.9 *Let $\sum a_k$ be a real series with $a_k \geq 0$ for all k. Then $\sum a_k$ has a well-defined limit in $\mathbb{R}_\infty$ and converges in $\mathbb{R}$ if its sequence of partial sums is bounded.*

2.3 Absolute Convergence

A complex series $\sum z_k$ is said to converge *absolutely* if $\sum |z_k| < \infty$. The completeness of $\mathbb{C}$ implies the following result:

Exercise 2.10 Prove that an absolutely convergent complex series is convergent in $\mathbb{C}$.

In fact, this property that absolute convergence of a series implies convergence is another equivalent form of the completeness axiom.

Convergence that is not absolute is called *conditional*. The Cauchy criterion of Theorem 2.8 still applies to such cases, and this does lead to convergence results for certain cases.

Example 2.11 The *alternating harmonic* series is
$$1 - \frac{1}{2} + \frac{1}{3} - \frac{1}{4} + \cdots.$$
By pairing the terms and noting that

$$\frac{1}{k} - \frac{1}{k+1} \le 0,$$

we can deduce that

$$\left| \frac{1}{n} - \frac{1}{n+1} + \cdots \pm \frac{1}{m} \right| \le \frac{1}{n},$$

where the $\pm$ depends on whether $m - n$ is odd or even. This shows that the alternating harmonic series satisfies the Cauchy criterion and is therefore convergent. $\diamond$

The argument from in Exercise 2.11 can be applied to more general alternating series, but this is still a rather special case. Conditional convergence is quite difficult to analyze in general.

One of the key features that makes absolute convergence much more straightforward than conditional is the ability to reorder the terms without changing the limit. A *rearrangement* of a series $\sum z_k$ is a new series given by $\sum z_{\sigma_k}$, where σ is a bijection $\mathbb{N} \to \mathbb{N}$ written as $k \mapsto \sigma_k$. Rearrangement alters the sequence of partial sums, and so it may affect the convergence properties in general. However, for an absolutely convergent series this is not an issue.

Theorem 2.12 *For an absolutely convergent series, all rearrangements converge to the same value.*

Proof Suppose $\sum z_k$ is absolutely convergent, with partial sums s_n. For the rearrangement defined by $\sigma : \mathbb{N} \to \mathbb{N}$, we have the new partial sums

$$s'_n := z_{\sigma_1} + \cdots + z_{\sigma_n}.$$

Given $\varepsilon > 0$, we will show that there exists an integer M so that

$$|s'_n - s_n| \le \varepsilon \tag{2.3}$$

for all $n \ge M$. Since ε is arbitrary, this implies that (s'_n) converges to the same limit as (s_n).

To establish (2.3), first choose m such that

$$\sum_{k=m}^{\infty} |z_k| \le \varepsilon, \tag{2.4}$$

which is possible by absolute convergence. Then set

$$N := \max(k : \sigma_k \le m),$$

which is finite because σ is invertible. This ensures that

$$\{1, \ldots, m\} \subset \{\sigma_1, \ldots, \sigma_N\}.$$

2.3 Absolute Convergence

Hence, for $n \geq N$, the difference $s'_n - s_n$ consists of a finite sum of terms a_k with $k > m$. By (2.4), this implies that $|s'_n - s_n| \leq \varepsilon$ for all $n \geq N$. □

Rearrangements of conditionally convergence series can change the limit. Indeed, a classic theorem of Riemann says that a real series that is conditionally convergent can be rearranged to converge to any real number, or to diverge to either $\pm\infty$.

Exercise 2.13 Consider two absolutely convergent series,

$$\sum_{k=0}^{\infty} a_k = A, \quad \sum_{k=0}^{\infty} b_k = B.$$

Prove that the series $\sum c_n$, where

$$c_n := \sum_{k=0}^{n} a_k b_{n-k},$$

converges absolutely with

$$\sum_{n=0}^{\infty} c_n = AB.$$

2.3.1 Convergence Tests

The standard method for establishing absolute convergence is through comparison to known series, using the following result:

Exercise 2.14 For some integer N suppose that $|z_k| \leq b_k$ for all $k \geq N$. Prove that if $\sum b_k < \infty$, then $\sum z_k$ converges absolutely.

Example 2.15 The *exponential* function is defined by the series

$$\exp(z) := \sum_{k=0}^{\infty} \frac{z^k}{k!}. \tag{2.5}$$

To check convergence using Exercise 2.14, a crude lower bound on the factorial will suffice. For $k \geq N$,

$$\frac{1}{k!} = \frac{1}{k(k-1)\cdots(N+1)N!}$$
$$\leq \frac{1}{N^{k-N} N!}.$$

Hence

$$\frac{|z|^k}{k!} \leq C_N \left|\frac{z}{N}\right|^k. \tag{2.6}$$

Comparison to the geometric series then implies that the exponential series (2.5) is absolutely convergent for $|z| < N$. Since N was arbitrary, this demonstrates absolute convergence for all $z \in \mathbb{C}$. ◊

The geometric series can be used more generally as a basis for comparison. For example, relating the terms z_k to a power r^k leads to the following:

Theorem 2.16 (root test) *For $z_k \in \mathbb{C}$ define the quantity*

$$q := \limsup_{k \to \infty} |z_k|^{1/k}. \tag{2.7}$$

The series $\sum z_k$ converges absolutely if $q < 1$ and diverges if $q > 1$.

Proof Suppose $q < 1$ and choose r so that $q < r < 1$. By the characterization of the limsup in Exercise 1.16, there exists N so that

$$|z_k|^{1/k} < r$$

for all $k \geq N$. Since this means $|z_k| \leq r^k$ and $r < 1$, Exercise 2.14 implies that $\sum z_k$ converges absolutely.

Now assume that $q > 1$. Then by Exercise 1.16 there are infinitely many k such that $|z_k| \geq 1$. This guarantees divergence by Exercise 2.4. □

The root test is inconclusive if $q = 1$. It also has the limitation that the formula (2.7) may be difficult to evaluate, because of the root. Another way to handle the comparison to geometric series is through the ratios of successive elements.

Exercise 2.17 (ratio test) Prove that the series $\sum z_k$ is absolutely convergent if

$$\limsup_{k \to \infty} \left|\frac{z_{k+1}}{z_k}\right| < 1.$$

For example, the ratio test for the exponential series reduces to the calculation

$$\limsup_{n \to \infty} \frac{|z|}{n+1} = 0,$$

which confirms the absolute convergence for all $z \in \mathbb{C}$.

2.4 Power Series

A complex *power series* is a function of $z \in \mathbb{C}$ given by

$$f(z) = \sum_{k=0}^{\infty} c_k (z - z_0)^k, \qquad (2.8)$$

with z_0 and c_n in $\mathbb{C}$. We will frequently set the center point $z_0 = 0$ for convenience, because this constant does not affect convergence. We have already seen in Example 2.6 the geometric power series,

$$\frac{1}{1-z} = \sum_{k=0}^{\infty} z^k, \quad \text{for } |z| < 1.$$

And the exponential power series was introduced in Example 2.15,

$$\exp(z) := \sum_{k=0}^{\infty} \frac{z^n}{n!}, \quad \text{for all } z \in \mathbb{C}.$$

For a general power series of the form (2.8), the root test gives a straightforward convergence result. In this case, the quantity q in the root test formula (2.7) reduces to $|z - z_0|/R$, where R is defined as the *radius of convergence* of the series,

$$R := \frac{1}{\limsup |c_k|^{1/k}}, \qquad (2.9)$$

with the interpretation that $R = \infty$ if the denominator is 0 and $R = 0$ if the denominator is infinite. From Theorem 2.16 we have the following:

Theorem 2.18 *The power series (2.8) converges absolutely for $|z - z_0| < R$ and diverges for $|z - z_0| > R$.*

We have already seen examples of $R = 1$ for the geometric series and $R = \infty$ for the exponential series. If $R = 0$ then the series is effectively meaningless, so we always assume that $R > 0$ for a general power series.

A natural question to ask at this point is why powers of $\bar{z}$ are not included in the definition of complex power series. The answer has to do with existence of the complex derivative with respect to z, which fails if any $\bar{z}$ terms are included. This issue falls within the realm of complex analysis, which we will not go further into here.

The constant $e := \exp(1)$ is known as *Euler's number*. For real variables it is possible to define the power e^x independently, by first considering rational powers, and then to prove that $e^x = \exp(x)$. For complex numbers, we simply take the exponential series as a definition of the power,

$$e^z := \exp(z) \qquad (2.10)$$

for $z \in \mathbb{C}$. The following addition formula shows that this is consistent with definitions in the real or rational cases.

Lemma 2.19 *For $z, w \in \mathbb{C}$,*

$$\exp(z + w) = \exp(z)\exp(w). \tag{2.11}$$

Proof By the product formula from Exercise 2.13,

$$\exp(z)\exp(w) = \sum_{n=0}^{\infty} \sum_{k=0}^{n} \frac{z^k}{k!} \frac{w^{n-k}}{(n-k)!}.$$

The sum over k can be evaluated using the binomial formula,

$$\sum_{k=0}^{n} \frac{z^k}{k!} \frac{w^{n-k}}{(n-k)!} = \frac{(z+w)^n}{n!},$$

which yields the result. $\square$

Trigonometric functions can also be defined for complex arguments via power series. For example,

$$\cos(z) := 1 - \frac{z^2}{2!} + \frac{z^4}{4!} - \ldots,$$

and

$$\sin(z) := z - \frac{z^3}{3!} + \frac{z^5}{5!} - \ldots,$$

both with radius $R = \infty$. By comparing these to the exponential series, we can immediately see that

$$\cos(z) = \frac{e^{iz} + e^{-iz}}{2}, \quad \sin(z) = \frac{e^{iz} - e^{-iz}}{2i}. \tag{2.12}$$

For $\theta \in \mathbb{R}$, solving for the exponential in (2.12) yields *Euler's formula* :

$$e^{i\theta} = \cos\theta + i\sin\theta.$$

By the addition formula (2.11),

$$|e^{i\theta}|^2 = e^{i\theta} \cdot \overline{e^{i\theta}} = e^{i\theta} e^{-i\theta} = 1,$$

which implies the trigonometric identity

$$\cos^2\theta + \sin^2\theta = 1.$$

2.4 Power Series

This shows that $(\cos\theta, \sin\theta)$ parametrizes the unit circle, and it is then relatively easy to show that the restrictions of the complex functions sin and cos to $\mathbb{R}$ agree with the original definitions in terms of the geometry of right triangles.

Euler's formula yields a very useful geometric interpretation of complex numbers. If (r, θ) is the polar coordinate representation of the Cartesian point (x, y), then $z = x + iy$ can be written as

$$z = re^{i\theta}.$$

In terms of the geometry of the plane, multiplication by z corresponds to dilation by r coupled with rotation by angle θ.

Metric Topology

To define limits for real or complex sequences we used the absolute value to measure the distance between points. The same limit definitions can be applied to any set that is equipped with a suitable distance function. From the notion of distance we can derive a collection of interrelated concepts that includes limits, continuity, compactness, etc. These are the basic notions of topology, and the term *metric topology* refers to the case where all of these concepts are derived from a distance function.

3.1 Metric Spaces

On a set X, a *distance function* (also called a *metric*) is a map

$$d(\cdot, \cdot) : X \times X \to [0, \infty)$$

satisfying the following properties for all points of X:

(i) symmetry: $d(x, y) = d(y, x)$;
(ii) definiteness: $d(x, y) = 0$ if and only if $x = y$;
(iii) triangle inequality: $d(x, y) \leq d(x, z) + d(z, y)$.

The combination (X, d) is called a *metric space*. We have already seen $\mathbb{R}$ and $\mathbb{C}$ as foundational examples, with

$$d(z, w) := |z - w|$$

in both cases.

Example 3.1 The space $\mathbb{R}^n$ is the set of vectors $v = (v_1, \ldots, v_n)$ with components $v_j \in \mathbb{R}$. The standard metric on this space is derived from the *Euclidean norm*,

$$\|v\| := \sqrt{v \cdot v} = \sqrt{v_1^2 + \cdots + v_n^2}. \tag{3.1}$$

The associated distance function is

$$d(v, w) := \|v - w\|. \tag{3.2}$$

The symmetry and positive definiteness of (3.2) are obvious from the definition. The triangle inequality in this case follows from the vector identity

$$\|u + v\|^2 = \|u\|^2 + 2\|u\|\|v\| \cos \theta + \|v\|^2,$$

where θ is the angle between u and v. ◇

Example 3.2 The extended real number system $\mathbb{R}_\infty$ introduced in Sect. 1.2 can be interpreted as a metric space using the stereographic projection illustrated in Fig. 3.1. This map identifies the angle $\theta \in (-\frac{\pi}{2}, \frac{\pi}{2})$ on the half-circle with the point $\tan \theta \in \mathbb{R}$. We extend this to a bijective map $[-\frac{\pi}{2}, \frac{\pi}{2}] \to \mathbb{R}_\infty$ by defining

$$\tan(\pm \tfrac{\pi}{2}) := \pm \infty.$$

The distance on $\mathbb{R}_\infty$ can then be measured as the difference in angle on the semicircle. This gives a distance function

$$d_\infty(x, y) := |\arctan x - \arctan y|.$$

The triangle inequality for d_∞ follows from the ordinary triangle inequality on $[-\pi/2, \pi/2]$. ◇

The Euclidean space discussed in Example 3.1 is an example of a very common type of metric space, the *normed vector space*. A *norm* $\|\cdot\|$ on a vector space V is a function $V \to [0, \infty)$ which satisfies:

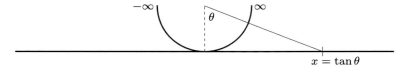

Fig. 3.1 The real axis is identified with the semicircle by stereographic projection, with $\pm \infty$ mapping to the endpoints. The metric d_∞ metric corresponds to arclength on the semicircle

3.1 Metric Spaces

(i) *homogeneity*: $\|cv\| = |c|\|v\|$ for any scalar c;
(ii) *definiteness*: $\|v\| = 0$ if and only if 0;
(iii) *triangle inequality*: $\|u + v\| \leq \|u\| + \|v\|$.

Comparing these properties to the definition of distance makes it clear that a normed vector space is a metric space with the default distance function,

$$d(v, w) := \|v - w\|.$$

Example 3.3 An alternative to the Euclidean norm (3.1) on $\mathbb{R}^n$ is given by the *max norm*,

$$\|v\|_{\max} := \max_j |v_j|.$$

for $v = (v_1, \ldots, v_n) \in \mathbb{R}^n$. The two norms are related by inequalities:

$$\|v\|_{\max} \leq \|v\| \leq \sqrt{n}\|v\|_{\max} \tag{3.3}$$

for all v. $\diamond$

Example 3.4 For $p \geq 1$ consider the space of p-summable sequences in $\mathbb{C}$, which is denoted

$$\ell^p := \left\{\alpha = (\alpha_1, \alpha_2, \ldots) : \sum_{j=1}^{\infty} |\alpha_j|^p < \infty\right\},$$

equipped with the norm

$$\|\alpha\|_p := \left(\sum_{j=1}^{\infty} |\alpha_j|^p\right)^{1/p}.$$

The power of $1/p$ guarantees that the homogeneity condition is satisfied, and positive definiteness is obvious. This leaves the triangle inequality to check. The ℓ^p version of the triangle inequality is called the *Minkowski inequality*:

$$\|\alpha + \gamma\|_p \leq \|\alpha\|_p + \|\gamma\|_p. \tag{3.4}$$

For $p = 1$ the Minkowski inequality is a simple extension of the complex triangle inequality, which gives

$$\sum_{j=1}^{n} |\alpha_j + \beta_j| \leq \sum_{j=1}^{n} |\alpha_j| + \sum_{j=1}^{n} |\beta_j|$$

for any finite n. Since the terms of these sums are positive, Theorem 2.9 allows us to take $n \to \infty$ to conclude that

$$\|\alpha + \beta\|_1 \leq \|\alpha\|_1 + \|\beta\|_1.$$

This proves the $p = 1$ triangle inequality and also confirms that ℓ^1 is a vector space.

The Minkowski inequality is a bit trickier to prove for $p > 1$, but the details involve some estimates that are very useful in their own right. The first is *Young's inequality*: if $p > 1$ and $q > 1$ satisfy
$$\frac{1}{p} + \frac{1}{q} = 1,$$
then
$$xy \leq \frac{x^p}{p} + \frac{y^q}{q} \tag{3.5}$$
for all $x, y > 0$. We will defer the proof of Young's inequality until we have introduced the logarithm in Sect. 5.3.

The next step is *Hölder's inequality* for ℓ^p, which says that for $\alpha \in \ell^p$ and $\beta \in \ell^q$, with p, q as above,
$$\|\alpha\beta\|_1 \leq \|\alpha\|_p \|\beta\|_q, \tag{3.6}$$
where $\alpha\beta$ denotes the pointwise product. By the homogeneity of the norms, it suffices to prove that
$$\|\alpha\beta\|_1 \leq 1$$
for $\|\alpha\|_p = 1$ and $\|\beta\|_q = 1$. This follows from Young's inequality by setting $x = |\alpha_j|$ and $y = |\beta_j|$ in (3.5) and then summing over j.

Finally, we note that (3.6) implies that for $\alpha \in \ell^p$ and $\beta \neq 0 \in \ell^q$, with p and q as above,
$$\|\alpha\|_p \geq \frac{\|\alpha\beta\|_1}{\|\beta\|_q}. \tag{3.7}$$
Since $(p-1)q = p$, the sequence $|\alpha|^{p-1}$ lies in ℓ^q, and setting $\beta = |\alpha|^{p-1}$ yields an equality in (3.7). Therefore, Hölder's inequality implies that
$$\|\alpha\|_p = \sup_{\|\beta\|_q = 1} \|\alpha\beta\|_1, \tag{3.8}$$
for $p > 1$ and $q = p/(p-1)$. This is useful because we already checked the triangle inequality for the ℓ^1 norm. For $\alpha, \beta \in \ell^p$,
$$\begin{aligned}\|\alpha + \beta\|_p &= \sup_{\|\gamma\|_q=1} \|(\alpha+\beta)\gamma\|_1 \\ &\leq \sup_{\|\gamma\|_q=1} \left(\|\alpha\gamma\|_1 + \|\beta\gamma\|_1\right) \\ &\leq \sup_{\|\gamma\|_q=1} \|\alpha\gamma\|_1 + \sup_{\|\gamma\|_q=1} \|\beta\gamma\|_1 \\ &= \|\alpha\|_p + \|\beta\|_p.\end{aligned}$$
This completes the proof of (3.4). ◊

3.2 Open and Closed Sets

Exercise 3.5 The definition given in Example 3.4 can be extended to $p = \infty$ by setting
$$\ell^\infty := \{\alpha = (\alpha_1, \alpha_2, \dots) : \sup |\alpha_j| < \infty\},$$
with the norm
$$\|\alpha\|_\infty := \sup_j |\alpha_j|.$$
Justify the notation by proving that for $\alpha \in \ell^1$,
$$\|\alpha\|_\infty = \lim_{p \to \infty} \|\alpha\|_p.$$

3.2 Open and Closed Sets

The *neighborhood* of a point $x \in X$ with radius $r > 0$ is defined as
$$N_r(x) := \{y \in X : d(x, y) < r\}.$$

For example, in Euclidean $\mathbb{R}^n$ the neighborhood $N_r(x)$ is the interior of a spherical ball of radius r centered at x. For $n = 1$ this reduces to an interval centered at x.

Example 3.6 The family of ℓ^p norms defined in Example 3.4 and Exercise 3.5 can be applied to $\mathbb{R}^n$ as well, by regarding a point $x = (x_1, \dots, x_n)$ as a finite sequence. For $p \in [1, \infty)$ this gives
$$\|x\|_p := \left(x_1^p + \cdots + x_n^p\right)^{1/p},$$
while $\|\cdot\|_\infty$ is equal to the max norm (4.11). Of course, $\|\cdot\|_2$ is the familiar Euclidean norm. Some neighborhoods for different values of p are illustrated in Fig. 3.2.

The concept of a neighborhood is used to define various classes of points in relation to a subset. Given a subset $E \subset X$:

Fig. 3.2 Neighborhoods in $\mathbb{R}^2$ with respect to ℓ^p norms

(i) $x \in E$ is an *interior point* if $N_\varepsilon(x) \subset E$ for some $\varepsilon > 0$.
(ii) $x \in E$ is an *isolated point* if $N_\varepsilon(x) \cap E = \{x\}$ for some $\varepsilon > 0$.
(iii) $x \in X$ is a *limit point* of E if every neighborhood $N_\varepsilon(x)$ contains at least one point $y \in E$ with $y \neq x$.
(iv) x is a *boundary point* of E if every neighborhood $N_\varepsilon(x)$ intersects both E and E^c.

Note that interior and isolated points are required to be elements of E, while limit and boundary points need not be. Some special notations are used for these classes,

$$E^\circ := \{\text{interior points of } E\},$$
$$E' := \{\text{limit points of } E\},$$
$$\partial E := \{\text{boundary points of } E\}.$$

Example 3.7 Consider $\mathbb{R}$ as a metric space. For the open interval $I = (a, b)$ the set of limit points is the closed interval $[a, b]$. All points of (a, b) are interior and there are no isolated points. The endpoints of the interval are boundary points.

For the subset of integers, all points of $\mathbb{Z}$ are isolated boundary points. There are no limit points or interior points in this case.

For $\mathbb{Q}$ as a subset of $\mathbb{R}$, every real number qualifies as a limit point and a boundary point, by Theorem 1.3. There are no interior or isolated points. ◊

Here are some exercises to reinforce these fundamental definitions:

Exercise 3.8 Decide whether the statements below are true or false, and either give a proof or find a counterexample. All statements refer to a subset E of a metric space X.

(a) Every point in E set is either interior or isolated.
(b) Every point in E is either interior or boundary.
(c) All points of X are either interior to E, interior to E^c, or boundary points.
(d) A point of E that is not isolated is a limit point.
(e) A limit point of E is either interior or a boundary point.

Exercise 3.9 A point $p \in X$ is called an *accumulation point* of $E \subset X$ if every neighborhood of p contains infinitely many points of E. Prove that the definitions of accumulation point and limit point are equivalent.

From the classification of points introduced above, we obtain the fundamental definitions of metric topology. For a subset $E \subset X$,

(i) E is *open* if all its points are interior.
(ii) E is *closed* if it contains all of its limit points.

3.2 Open and Closed Sets

This definition is consistent with the terminology of open and closed intervals in $\mathbb{R}$. A set may be neither open nor closed, as the example of a half-open interval $(a, b]$ demonstrates. On the other hand, the empty set is both open and closed, as is $\mathbb{R}$ itself. In any metric space X, both of the sets $\emptyset$ and X automatically qualify as both open and closed.

Exercise 3.10 Prove that neighborhoods are open.

Exercise 3.11 Prove that a set is open if and only if its complement is closed.

The behavior of open and closed sets under complements and other basic set operations is described in the following:

Exercise 3.12 Prove the following:

(a) The union of any collection of open sets is open.
(b) The intersection of a finite collection of open sets is open.
(c) The intersection of any collection (not necessarily countable) of closed sets is closed.
(d) The union of a finite collection of closed sets is closed.

The interior A° of a set $A \in X$ is often described as the *largest* open set contained in A. This means that if $U \subset A$ is open, then $U \subset A^\circ$. This follows immediately from the definition of interior point. Since the union of an arbitrary collection of open sets is open by Exercise 3.12, we could also characterize A° as the union of all open subsets of A.

We can similarly define the *closure* of A, denoted by $\overline{A}$, as the smallest closed set containing A, or equivalently as the intersection of all closed sets containing A. The following result gives another characterization of the closure which is often used as the definition.

Theorem 3.13 *For a subset $A \subset X$,*

$$\overline{A} = A \cup A',$$

where A' is the set of limits points of A.

Proof Let $B = A \cup A'$. Our first claim is that B is closed. To prove this we will show that B^c is open. If $x \notin B$, then there exists a neighborhood $N_\varepsilon(x)$ that contains no point of A other than possibly x. If $x \notin A$ also, then $N_\varepsilon(x) \cap A = \emptyset$. Because $N_\varepsilon(x)$ is open, this implies that $N_\varepsilon(x)$ contains no limit points of A either. Hence $N_\varepsilon(x) \cap B = \emptyset$ and so $N_\varepsilon(x) \subset B^c$. This proves that B^c is open, and hence B is closed.

Now suppose that F is closed and $A \subset F$. A limit point of A is also a limit point of F, by definition, which means that $A' \subset F$ also. Thus $B \subset F$, and therefore B is the smallest closed set containing A. $\square$

Exercise 3.14 Show that
$$\overline{A} = A \cup \partial A,$$
where ∂A is the set of boundary points of A.

Exercise 3.15 Determine if these statements are true or false, and give either a proof or counterexample:

(a) $\overline{A \cup B} = \overline{A} \cup \overline{B}$
(b) $\overline{A \cap B} = \overline{A} \cap \overline{B}$
(c) $(A \cup B)^\circ = A^\circ \cup B^\circ$
(d) $(A \cap B)^\circ = A^\circ \cap B^\circ$

We say that a subset E is *dense* in X if $\overline{E} = X$. For example, Theorem 1.3 implies that $\mathbb{Q}$ is dense in $\mathbb{R}$.

Exercise 3.16 Prove that a set E is dense in X if and only if E intersects every non-empty open set in X.

3.2.1 Subspace Topology

Given a metric space (X, d) and a subset $Y \subset X$, we can use d to measure distance even when our attention is restricted to points of Y. The restriction $d|_Y$ clearly satisfies the requirements for a distance function on Y, and thus Y inherits a natural metric structure from the larger set. To indicate that the metric was obtained in this way, we call Y a *metric subspace* of X. The resulting definitions of interior points, limit points, open and closed sets, etc., are collectively referred to as the *subspace topology* on Y.

When using subspace topology, we need to be very careful about the context of topological terms, because they can have a different meaning in the two topologies. For example, suppose $Y = [0, 1]$ is defined as a metric subspace of $\mathbb{R}$. Then a neighborhood of the point 1 in Y has the form $N_\varepsilon(1) = (1 - \varepsilon, 1]$, which does not qualify as a neighborhood in $\mathbb{R}$. The set Y is open in the subspace topology but not in the topology of $\mathbb{R}$.

Given a fixed metric space X, the default assumption is that all terms refer to the topology of X unless otherwise indicated. The qualification "relative to Y" is added to specify the use of the subspace topology.

Theorem 3.17 *Let X be a metric space and $Y \subset X$. A set $E \subset Y$ is open relative to Y if and only if there is an open set $G \subset X$ such that*
$$E = G \cap Y.$$

3.2 Open and Closed Sets

Proof Let us use $N_r(x)$ to denote a neighborhood in X, and $V_r(y)$ for a neighborhood relative to Y. Since the distance function on Y is defined by restriction,

$$V_r(y) = N_r(y) \cap Y \tag{3.9}$$

for $y \in Y$.

If $G \subset X$ is open, then for each $y \in G \cap Y$, we can find $\varepsilon > 0$ such that $N_\varepsilon(y) \subset G$. By (3.9), the corresponding neighborhood $V_\varepsilon(y)$ is contained in $G \cap Y$. Therefore $G \cap Y$ is open relative to Y.

Now assume that $E \subset Y$ is open relative to Y. For each $y \in E$, we find $r_y > 0$ such that $V_{r_y}(x) \subset E$. If we then define G by setting

$$G := \bigcup_{y \in E} N_{r_y}(y),$$

then G is open by Exercise 3.12, and (3.9) implies that

$$G \cap Y = \bigcup_{y \in E} V_{r_y}(y) = E.$$

$\square$

The subspace topology allows us to give a concise formulation of connectedness without making any reference to paths between points. A subset $E \subset X$ is defined to be *connected* if the only subsets of E which are both open and closed relative to E are $\emptyset$ and E.

We can make this definition a bit more intuitive by introducing a complementary notion. Two non-empty subsets A and B are *separated* if both $\overline{A} \cap B$ and $A \cap \overline{B}$ are empty.

Exercise 3.18 Prove that a subset is connected if and only if it cannot be written as the union of two non-empty separated sets.

If A and B are disjoint non-empty closed sets, then the requirements for separation are automatically satisfied, because $\overline{A} = A$ and $\overline{B} = B$. The same conclusion holds for disjoint open sets.

Exercise 3.19 Prove that two non-empty open sets are separated if they are disjoint.

We defined an *interval* in Sect. 1.1 as a convex subset of the real numbers. The following result shows that connectedness in $\mathbb{R}$ is equivalent to convexity.

Theorem 3.20 *The connected subsets of $\mathbb{R}$ are precisely the intervals.*

Proof To see that an interval $I \subset \mathbb{R}$ is connected, suppose that $I = A \cap B$, where A and B are disjoint and not empty. Pick $a \in A$ and $b \in B$. We can assume that $a < b$ by switching the sets if necessary. Then let
$$x = \sup(A \cap [a, b]).$$
Since I is an interval, all points in $(x, b]$ are contained in I and must therefore be elements of B. This implies $x \in \overline{B}$. We also know that $x \in \overline{A}$ by Exercise 1.7. Since $x \in A \cup B$, this shows that A and B are not separated. We conclude that I is connected.

Now assume $E \subset \mathbb{R}$ is connected. To prove that E is an interval we suppose that $x, y \in E$ with $x < y$. For $x < t < y$, let $A = (-\infty, t) \cap E$ and $B = (t, \infty) \cap E$. These sets are nonempty and separated, so it follows from Exercise 3.18 that $E \neq A \cup B$. This shows that $t \in E$ for all $t \in (x, y)$, which proves that E is an interval. □

3.3 Convergence and Completeness

In a metric space (X, d), a sequence (x_k) *converges* to y if
$$\lim_{k \to \infty} d(x_k, y) = 0.$$
This clearly generalizes the definition given in Sect. 1.2 for real numbers, and we use the same abbreviated notations when the index is clear: $\lim x_k = y$ or $x_k \to y$.

Exercise 3.21 Prove that x_0 is a limit point of $E \subset X$ if and only if there is a sequence (x_n) in $E \setminus \{x_0\}$ such that $x_n \to x_0$.

Exercise 3.22 For $x_n \in \mathbb{R}$ and $\alpha \in \mathbb{R}_\infty$, prove that $x_n \to \alpha$ in the metric space $(\mathbb{R}_\infty, d_\infty)$ defined in Example 3.2 if and only if $x_n \to \alpha$ in the extended sense defined in Sect. 1.2.

A sequence (x_n) is *Cauchy* if for every $\varepsilon > 0$, there exists N such that
$$d(x_m, x_n) < \varepsilon \quad \text{for all } m, n \geq N. \tag{3.10}$$
Convergent sequences are Cauchy by the triangle inequality, as we observed in Lemma 1.22. The converse is not necessarily true, however. For example, consider $X = (0, 1)$ as a subspace of $\mathbb{R}$. A sequence in $(0, 1)$ that converges to 0 in $\mathbb{R}$ is Cauchy but not convergent in X.

A Cauchy sequence (x_n) is necessarily *bounded*, which means there exists $R > 0$ such that
$$d(x_n, x_m) \leq R$$

3.3 Convergence and Completeness

for all n, m. This is clear because all but finitely many points lie within a neighborhood of finite radius. Another useful property of Cauchy sequences is the following result, which is the generalization of the argument used in Exercise 1.26.

Exercise 3.23 Suppose (x_n) is a Cauchy sequence. Prove that if there is a subsequence (x_{n_k}) such that
$$\lim_{k \to \infty} x_{n_k} = x,$$
then the full sequence converges to x.

A metric space is *complete* if every Cauchy sequence converges. This proves to be a crucial property for most applications of metric spaces. Indeed, we do not normally consider a metric space to be properly defined unless it is complete. This is because we often need to establish the convergence of a sequence whose limit is unknown, and without the Cauchy criterion this may not be possible.

For metric spaces defined using a subspace topology, as described in Sect. 3.2.1, the issue of completeness is easily resolved if the ambient space is already known to be complete.

Exercise 3.24 Suppose X is a complete metric space. For a subset $Y \subset X$, prove that Y is complete as a metric subspace if and only if Y is closed.

We have already seen that $\mathbb{R}$ is metrically complete in Theorem 1.23, and $\mathbb{C}$ in Exercise 2.3. These results are easily extended to Euclidean vector spaces of higher dimension.

Theorem 3.25 *The Euclidean vector spaces $\mathbb{R}^n$ and $\mathbb{C}^n$ are complete.*

Proof It suffices to prove the real case, since $\mathbb{C}^n$ is equivalent to $\mathbb{R}^{2n}$ under the Euclidean norm. We have already shown that $\mathbb{R}$ is metrically complete in Theorem 1.23.

Let (v_k) be a bounded sequence in $\mathbb{R}^n$, with v_k written in coordinates as $(v_{k,1}, \ldots, v_{k,n})$. Since
$$\|v_k - v_m\| \geq |v_{k,j} - v_{m,j}|$$
for each j, each sequence of coordinates $(v_{k,j})_{k \in \mathbb{N}}$ is Cauchy in $\mathbb{R}$. By Theorem 1.23 the limits
$$w_j := \lim_{k \to \infty} v_{k,j}$$
exist for each j. This defines a vector $w \in \mathbb{R}^n$. We have $v_k \to w$ because
$$\|v_k - w\|^2 = \sum_{j=1}^{n} |v_{k,j} - w_j|^2,$$
and the terms in the sum all approach 0 as $k \to \infty$. □

Although Theorem 3.25 refers specifically to the Euclidean metric, it turns out that all finite-dimensional normed vector spaces are complete, for any choice of norm. To establish this fact, we will show that all finite-dimensional norms generate the same topology. For this purpose, it will be helpful to first extend the Bolzano-Weierstrass property (Theorem 1.25) to higher dimension.

Theorem 3.26 (Bolzano-Weierstrass) *A bounded sequence in $\mathbb{R}^n$ or $\mathbb{C}^n$ has a convergent subsequence.*

Proof As above, it suffices to consider the real case. Let (v_k) be a bounded sequence in $\mathbb{R}^n$, written in coordinates as $(v_{k,1}, \ldots, v_{k,n})$. Since each sequence $(v_{k,1})_{k=1}^{\infty}$ is bounded in $\mathbb{R}$, Theorem 1.25 implies that by passing to a subsequence we can assume that $v_{k,1} \to w_1 \in \mathbb{R}$. We can then passing to a further subsequence to obtain $v_{k,2} \to w_2 \in \mathbb{R}$, and so on. After n steps, we will have reduced the original sequence to a subsequence such that converges to w in $\mathbb{R}^n$. □

Cauchy sequences are bounded, as noted above. Thus, a metric space that has the Bolzano-Weierstrass property is complete by Exercise 3.23. Not all complete metric spaces have the Bolzano-Weierstrass property, however, as we will see below.

On a vector space V, the norms $\|\cdot\|_1$ and $\|\cdot\|_2$ are considered *equivalent* if there exists constants $c, C > 0$ such that

$$c\|v\|_2 \le \|v\|_1 \le C\|v\|_2 \tag{3.11}$$

for all $v \in V$. For equivalent norms, (3.11) implies that any neighborhood defined with respect to one of the norms contains a neighborhood defined with respect to the other. This means that the definitions of all open and closed sets and all other topological notions from Sect. 3.2 will be identical for the two norms. In other words, equivalent norms generate the same topology. In particular, they yield the same definitions of convergence and of Cauchy sequences.

We will demonstrate the equivalence of any two norms on a finite-dimensional vector space as an application of the Bolzano-Weierstrass theorem.

Theorem 3.27 *On a finite-dimensional real or complex vector space, all norms are equivalent.*

Proof It suffices to prove this for $\mathbb{R}^n$ or $\mathbb{C}^n$, and since the real and complex cases have essentially the same argument, we consider only the real case.

Let $\|\cdot\|$ be the Euclidean norm on $\mathbb{R}^n$, and let $\|\cdot\|_*$ denote some other norm. Writing $x = (x_1, \ldots, x_n) \in \mathbb{R}^n$ as $\sum_{j=1}^n x_j e_j$ with respect to the standard basis $\{e_j\}$ and applying the triangle inequality for $\|\cdot\|_*$ gives

3.3 Convergence and Completeness

$$\|x\|_* \leq \sum_{j=1}^{n} |x_j| \|e_j\|_*$$
$$\leq n \max_j(|x_j|) \max_j(\|e_j\|_*).$$

Since $|x_j| \leq \|x\|$ for each j, this yields

$$\|x\|_* \leq c_1 \|x\|, \tag{3.12}$$

with $c_1 > 0$.

To prove the reverse inequality, suppose for the sake of contradiction that there is no constant C such that $\|v\| \leq C\|v\|_*$ for all $v \in \mathbb{R}^n$. Then for each $k \in \mathbb{N}$ we can find a vector $v_k \in \mathbb{R}^n$ such that $\|v_k\| = 1$ and

$$\|v_k\|_* < \frac{1}{k}. \tag{3.13}$$

Since the v_k are Euclidean unit vectors, Theorem 3.26 implies that a subsequence (v_{k_n}) converges to some $w \in \mathbb{R}^n$ in the Euclidean sense,

$$\|v_{k_n} - w\| \to 0. \tag{3.14}$$

Clearly $w \neq 0$, since $\|v_{k_n}\| = 1$. On the other hand, (3.12) and (3.14) imply that $\|v_k - w\|_* \to 0$ also. This gives $w = 0$ by (3.13), a contradiction. This proves that there exists a constant $c_2 > 0$ such that

$$\|x\| \leq c_2 \|x\|_* \tag{3.15}$$

for all $x \in \mathbb{R}^n$.

The inequalities (3.12) and (3.15) demonstrate that $\|\cdot\|_*$ is equivalent to the Euclidean norm. Hence any two norms on $\mathbb{R}^n$ are equivalent to each other. □

Theorems 3.26 and 3.27 yield the following:

Corollary 3.28 *Every finite-dimensional normed vector space is complete and has the Bolzano-Weierstrass property.*

The story is quite different for infinite-dimensional normed vector spaces, where completeness is not guaranteed and the Bolzano-Weierstrass property may fail even when spaces are complete.

Example 3.29 We will show that ℓ^p is complete for $1 \leq p < \infty$ but does not satisfy Bolzano-Weierstrass. Suppose that (α_k) is a Cauchy sequence in ℓ^p. For convenience let us notate each sequence as a map

$$\alpha_k : \mathbb{N} \to \mathbb{C}.$$

The fact that (α_k) is Cauchy implies that the sequences $(\alpha_k(j))_{k=1}^\infty$ are Cauchy in $\mathbb{C}$ for each j. Hence, by Theorem 3.26 we can define $\beta : \mathbb{N} \to \mathbb{C}$ by taking the limits

$$\beta(j) := \lim_{k \to \infty} \alpha_k(j).$$

Our goal is to show that $\beta \in \ell^p$ and that $\alpha_k \to \beta$ with respect to the p-norm.

Since Cauchy sequences are bounded, there exists $M > 0$ so that $\|\alpha_k\|_p \leq M$ for all k. For $n \in \mathbb{N}$, the Minkowski inequality (3.4) implies that

$$\left(\sum_{j=1}^n |\beta(j)|^p\right)^{1/p} \leq \left(\sum_{j=1}^n |\alpha_k(j) - \beta(j)|^p\right)^{1/p} + \left(\sum_{j=1}^n |\alpha_k(j)|^p\right)^{1/p}$$

$$\leq \left(\sum_{j=1}^n |\alpha_k(j) - \beta(j)|^p\right)^{1/p} + M.$$

Taking $k \to \infty$, with n fixed, yields

$$\left(\sum_{j=1}^n |\beta(j)|^p\right)^{1/p} \leq M.$$

We can then let $n \to \infty$ to prove that $\|\beta\|_p \leq M$.

To show that $\alpha_k \to \beta$, let $\varepsilon > 0$. By the Cauchy condition, there exists N so that $\|\alpha_k - \alpha_m\|_p \leq \varepsilon$ for $k, m \geq N$. For any $n \in \mathbb{N}$ this implies in particular that

$$\left(\sum_{j=1}^n |\alpha_k(j) - \alpha_m(j)|^p\right)^{1/p} \leq \varepsilon.$$

Taking $m \to \infty$ in this inequality gives

$$\left(\sum_{j=1}^n |\alpha_k(j) - \beta(j)|^p\right)^{1/p} \leq \varepsilon.$$

Since n was independent of ε, we can then let $n \to \infty$ to see that

$$\|\alpha_k - \beta\|_p \leq \varepsilon$$

for $k \geq N$.

To see that the Bolzano-Weierstrass property does not hold, consider the sequence

$$\alpha_k(j) = \begin{cases} 1, & j = k, \\ 0, & \text{otherwise}. \end{cases}$$

This sequence is bounded since $\|\alpha_k\|_p = 1$ for all k. But

3.3 Convergence and Completeness

$$\|\alpha_k - \alpha_m\|_p = 2^{1/p}$$

for all $k \neq m$, so a convergent subsequence is not possible. ◊

Exercise 3.30 Prove that the space ℓ^∞ introduced in Exercise 3.5 is complete.

In Sect. 2.3 we remarked that the absolute convergence property for series could be interpreted as a completeness axiom. This turns out to be true for a general normed vector space V. A series $\sum u_k$ in V *converges* to an element $w \in V$ if

$$\lim_{n \to \infty} \sum_{k=1}^{n} u_k = w. \tag{3.16}$$

The limit of partial sums is defined in the topology of V, so (3.16) means that

$$\lim_{n \to \infty} \left\| \sum_{k=1}^{n} u_k - w \right\| = 0. \tag{3.17}$$

The series is said to converge absolutely if $\sum \|u_k\| < \infty$.

Theorem 3.31 *A normed vector space is complete if and only if every absolutely convergent series is convergent.*

Proof Suppose $(V, \|\cdot\|)$ is a complete and let $\sum u_k$ be an absolutely convergent series. Define the partial sums

$$s_n := \sum_{k=1}^{n} u_k.$$

By the triangle inequalty,

$$|s_n - s_m| \leq \sum_{k=n+1}^{m} \|u_k\|$$

for $n < m$. Since this is bounded by $\sum_{k=n+1}^{\infty} \|u_k\|$, absolute convergence implies that the sequence (s_n) is Cauchy. Therefore (s_n) converges by completeness.

Now suppose that V has the absolute convergence property, and let (v_k) be a Cauchy sequence. For each $n \in \mathbb{N}$, choose k_n so that

$$\|v_i - v_j\| \leq 2^{-n}$$

for $i, j \geq k_n$. This implies in particular that

$$\sum \|v_{k_{j+1}} - v_{k_j}\| < \infty.$$

By assumption the series $\sum(v_{k_{j+1}} - v_{k_j})$ is convergent. Since the partial sums are given by

$$\sum_{j=1}^{n}(v_{k_{j+1}} - v_{k_j}) = v_{n+1} - v_1,$$

we conclude that the subsequence (v_{k_n}) is convergent. Hence the full sequence converges by Exercise 3.23. $\square$

3.3.1 Metric Completion

Given a metric space which is not complete, we can define a completion by the same strategy that Cantor used to construct the real numbers. This means that points in the completion are represented by equivalence classes of Cauchy sequences.

For sequences in (X, d) the equivalence relation is defined by

$$(x_k) \sim (y_k) \quad \text{if} \quad \lim_{k \to \infty} d(x_k, y_k) = 0.$$

This relation is clearly reflexive and symmetric, and transitivity follows from the triangle inequality. Let

$$X^* := \{\text{equivalence classes of Cauchy sequences in } X\}.$$

We will use the notation $[(x_k)]$ to indicate the class of the sequence (x_k). Note that there is a natural inclusion $X \to X^*$ given by

$$x \mapsto [(x)], \tag{3.18}$$

where (x) denotes the constant sequence.

Exercise 3.32 For any two Cauchy sequences (x_k) and (y_k) in X, prove that $\lim_{k \to \infty} d(x_k, y_k)$ exists in $\mathbb{R}$. Furthermore, show that this limit depends only on the equivalence classes of (x_k) and (y_k).

Exercise 3.32 allows us to make the following definition for the distance function on X^*,

$$d^*\big([(x_k)], [(y_k)]\big) := \lim_{k \to \infty} d(x_k, y_k).$$

Theorem 3.33 *The metric space* (X^*, d^*) *is complete.*

Proof Let (α_n) be a Cauchy sequence in X^*. For each α_n, choose a representative Cauchy sequence $(x_k^n)_{k=1}^{\infty}$ in X. The fact that (α_n) is Cauchy with respect to d^* means that for $\varepsilon > 0$ there exists N so that

3.3 Convergence and Completeness

$$m, n \geq N \implies \lim_{k \to \infty} d(x_k^n, x_k^m) \leq \varepsilon. \tag{3.19}$$

For each n, we can use the fact that (x_k^n) is Cauchy to choose N_n so that

$$i, j \geq N_n \implies d(x_i^n, x_j^n) \leq \frac{1}{n}.$$

Now consider the sequence y_k with elements

$$y_m := x_{N_m}^m.$$

To see that this sequence is Cauchy, let $\varepsilon > 0$ and choose N so that (3.19) holds. We can estimate, for $k \geq \max(N_n, N_m)$,

$$d(y_m, y_n) \leq d(y_m, x_k^m) + d(x_k^n, x_k^m) + d(x_k^m, y_m)$$
$$\leq \frac{1}{m} + d(x_k^n, x_k^m) + \frac{1}{n}.$$

Taking $k \to \infty$ by (3.19) yields

$$d(y_m, y_n) \leq \frac{1}{m} + \frac{1}{n} + d^*(\alpha_m, \alpha_n). \tag{3.20}$$

This implies that (y_m) is Cauchy in X, since (α_n) is Cauchy in X^*.

We claim that $\alpha_n \to \beta$ in X^*, where $\beta = [(y_m)]$. To see this, we need to estimate

$$d^*(\alpha_n, \beta) = \lim_{m \to \infty} d(x_m^n, y_m).$$

Observe that
$$d(x_m^n, y_m) \leq d(x_m^n, y_n) + d(y_n, y_m)$$
$$\leq d(x_m^n, y_n) + \frac{1}{m} + \frac{1}{n} + d^*(\alpha_m, \alpha_n),$$

by (3.20). For m sufficiently large, $d(x_m^n, y_n) \leq 1/n$, so taking $m \to \infty$ gives

$$d^*(\alpha_n, \beta) \leq \frac{2}{n} + \limsup_{m \to \infty} d^*(\alpha_m, \alpha_n).$$

The fact that the (α_n) is Cauchy implies that the right-hand side approaches 0 as $n \to \infty$. □

Exercise 3.34 Show that the inclusion of X in X^* gives a dense subset.

3.4 Compact Sets

Compactness is a topological concept that plays a fundamental role in many applications of metric spaces. We first discuss the standard topological definition, and later in the section we will give an alternative formulation in terms of sequential convergence.

A set $K \subset X$ is *compact* if every open cover of K has a finite subcover, where *open cover* means a collection of open sets $\{U_\alpha\}_{\alpha \in \mathcal{J}}$, such that

$$K \subset \bigcup_{\alpha \in \mathcal{J}} U_\alpha.$$

The index set $\mathcal{J}$ is arbitrary here in particular need not be countable. To say that a given open cover $\{U_\alpha\}$ has a finite subcover means that we can choose finitely many of the sets, say $U_1, \ldots, U_m$, such that

$$K \subset \bigcup_{j=1}^{m} U_j.$$

The empty set is compact by default, since no sets are needed to cover. A finite set is also clearly compact, because any cover can be reduced to a subcover by taking at most one set per point. To prove that a set is not compact is in some sense easier than proving compactness, because we need only give one example of a cover that admits no finite subcover.

Example 3.35 The Euclidean space $\mathbb{R}^n$ is not compact. To show this, consider a cover consisting of neighborhoods $\{N_r(0)\}_{r>0}$. Since these sets are nested, for any finite subcollection with radii $r_1 < \cdots < r_m$ we obtain

$$\bigcup_{j=1}^{m} N_{r_j}(0) = N_{r_m}(0).$$

Hence no finite subcover could contain $\mathbb{R}$.

Similarly, a neighborhood $N_a(x_0) \subset \mathbb{R}^n$ is not compact because the cover $\{N_r(x_0)\}_{0<r<a}$ admits no finite subcover. ◊

Example 3.36 Suppose $E \subset X$ is an infinite set with no limit points, so that E is closed but its points are isolated. Each $x \in E$ has a neighborhood which contains no other point of E. The collection of these neighborhoods yields an open cover with no finite subcover, so E is not compact. ◊

A subset E of a metric space is said to be *bounded* if its diameter is finite, meaning that

$$\sup\{d(x, y) : x, y \in E\} < 0.$$

3.4 Compact Sets

By the triangle inequality, E is bounded if and only if there exists a point $x_0 \in X$ and a finite radius R such that $E \subset N_R(x_0)$.

Theorem 3.37 *Compact sets are closed and bounded.*

Proof Suppose that $K \subset X$ is compact. For $x_0 \in X$ we can write

$$X = \bigcup_{r>0} N_r(x_0).$$

Thus $\{N_r(x_0)\}_{r>0}$ covers K in particular. The existence of a finite subcover implies that

$$K \subset N_R(x_0)$$

for some finite value of R. Hence K is bounded.

To show that K is closed, suppose $y \in K^c$. Since

$$\bigcup_{r>0} \{x \in X : d(x, y) > r\} = X \setminus \{y\},$$

the left-hand side yields an open cover for K. These sets are nested, so the existence of a finite subcover implies that

$$K \subset \{x \in X : d(x, y) > \delta\},$$

for some $\delta > 0$. In particular, $N_\delta(y) \subset K^c$. Since y was arbitrary this proves that K^c is open and hence K is closed. □

The converse of Theorem 3.37 is false in general. For example, let $X = (-1, 1)$ as a metric subspace of $\mathbb{R}$. Then X is closed in the subspace topology and has diameter 2. However, the argument from Example 3.35 shows that X is not compact.

This example highlights a major difference between compactness and the other topological properties introduced in Sect. 3.2. The definitions of open and closed change their meanings in a subspace topology. In particular, every subset of a metric space is both open and closed relative to its own topology. The following result shows that compactness is an intrinsic property.

Exercise 3.38 Let X be a metric space with subspace $Y \subset X$. Prove that a subset $K \subset Y$ is compact relative to Y if and only if K is compact relative to X.

Establishing the compactness of a given set directly from the definition can be challenging. We need to start with an arbitrary open cover and give some procedure by which it can be reduced to a finite collection. However, if the full metric space X is compact, then the condition is easy to verify for subsets.

Lemma 3.39 *If X is a compact metric space and $F \subset X$ is closed, then F is compact.*

Proof Suppose X is compact and $F \subset X$ is closed. If $\{U_\alpha\}$ is an open cover for F, then we can form an open cover of X by taking $\{U_\alpha\} \cup F^c$. Since X is compact, there exists a finite subcover of the form
$$X \subset F^c \cup U_1 \cup \cdots \cup U_m.$$
This implies in particular that
$$F \subset U_1 \cup \cdots \cup U_m,$$
proving that $\{U_\alpha\}$ admits a finite subcover for F. □

Although we have observed that the converse of Theorem 3.37 is not true in general, there is one important category of spaces for which it does hold.

Theorem 3.40 (Heine-Borel) *A subset of a finite-dimensional normed vector space is compact if and only if it is closed and bounded.*

Proof In Theorem 3.27 we proved that all norms on finite-dimensional vector spaces are topologically equivalent. Hence it suffices to prove the result for Euclidean $\mathbb{R}^n$. The fact that a compact set is closed and bounded was already established in Theorem 3.37.

Let $E \subset \mathbb{R}^n$ be closed and bounded. For convenience, we can place E inside a closed cube $Q \subset \mathbb{R}^n$ of side length l for l sufficiently large. It suffices to show that Q is compact, since the compactness of E will then follow from Lemma 3.39.

Assume that Q is not compact, which means there is an open cover $\{U_\alpha\}$ which admits no finite subcover. Consider a subdivision of Q into a union of 2^n closed cubes of side $l/2$. If each of these smaller cubes had a finite subcover, then Q would have a finite subcover also. Hence, at least one of least one of these inner cubes, call it Q_1, does not admit a finite subcover.

We can then subdivide Q_1 in the same way, and choose a cube Q_2 of side $l/4$ inside Q_1 with no finite subcover. Continuing this process produces a sequence of nested cubes
$$Q_1 \supset Q_2 \supset \ldots$$
where each Q_j has side width $l/2^j$ and no cube in the sequence admits a finite subcover drawn from $\{U_\alpha\}$.

Now form a sequence by choosing a point $x_j \in Q_j$ for each j. For $i, j \geq N$ we have
$$|x_i - x_j| \leq 2^{-N} l.$$
Thus (x_j) is a Cauchy sequence. By completeness (Theorem 3.25), the sequence converges and we can set $x = \lim x_j$. Since the cubes are closed, $x \in Q_j$ for all j.

3.4 Compact Sets

Choose one of the covering sets U_α so that $x \in U_\alpha$. Since U_α is open, the fact that the diameter of Q_j tends to zero implies that $Q_j \subset U_\alpha$ for j sufficiently large. This contradicts the fact that Q_j does not admit a finite subcover. From this contradiction we conclude that every open cover of Q admits a finite subcover, and hence Q is compact. □

The Heine-Borel theorem does not apply to infinite-dimensional normed vector spaces, as illustrated by the following:

Example 3.41 Consider the space ℓ^p introduced in Example 3.4. Let B be the closed unit ball,
$$B = \{\alpha \in \ell^p : \|\alpha\|_p \leq 1\}.$$
For $0 < \varepsilon < 1$ and $n \in \mathbb{N}$, define the open sets
$$U_n := \{\alpha \in \ell^p : |\alpha_k| < \varepsilon \text{ for all } k \geq n\}.$$
Then $\cup_{n=1}^\infty U_n = \ell^p$, so $\{U_n\}$ forms an open cover of B in particular. Because the sets are nested, the union over any finite subcover is equal to U_N for some integer N. Since U_N does not contain B for any N, B is not compact. ◊

The proof of Theorem 3.40 features a nesting argument which often turns out to be useful for arguments involving compactness. Note that the hypotheses are somewhat different in the following statement. We do not assume that X is complete, or that the diameters of the sets go to zero.

Exercise 3.42 (nested compact set property) Let $\{K_j\}$ be a sequence of compact sets which are nested,
$$K_1 \supset K_2 \supset \ldots,$$
and not empty. Prove that the intersection $\cap_{j=1}^\infty K_j$ contains at least one point.

3.4.1 Sequential Compactness

A set K is *sequentially compact* if every sequence in K has subsequence that converges in K. We have already seen an example of this property in the context of Euclidean spaces. A sequence in a compact set $K \subset \mathbb{R}^n$ has a subsequence converging in K by Bolzano-Weierstrass (Theorem 3.26). Therefore, compact subsets of $\mathbb{R}^n$ are sequentially compact.

In metric spaces sequential compactness turns out to be equivalent to compactness defined in terms of open covers. Before proving this, it is helpful to introduce yet another alternate definition. A set K is called *limit-point compact* if every infinite subset of K has a limit point inside the set.

Lemma 3.43 *For metric spaces, sequential compactness is equivalent to limit-point compactness.*

Proof Suppose K is sequentially compact and let $A \subset K$ be an infinite set. Then A contains a sequence (x_k) made up of distinct points. Sequential compactness means that there exists a subsequence converging to some $y \in K$. Hence y is a limit point of A.

Now assume that K is limit point compact and let (x_k) be a sequence in K. If some element of the sequence is repeated infinitely many times, the sequence admits a constant subsequence, which obviously converges. Thus it suffices to consider the case where each x_k is repeated at most finitely many times. This implies that $\{x_k\}$ is an infinite set and therefore has a limit point $y \in K$ by assumption. To form a convergent subsequence, choose x_{k_n} for each n so that $d(x_{k_n}, y) < 1/n$. $\square$

We start by proving that compactness implies sequential compactness. This actually remains true in a general (non-metric) topology. However, we will restrict our attention to the metric case to simplify the proof.

Theorem 3.44 *A compact set in a metric space is sequentially compact.*

Proof Suppose K is a compact subset of a metric space. By Lemma 3.43 it suffices to show that K is limit point compact.

Let $A \subset K$ be an infinite set and, for the sake of contradiction, assume that A has no limit point in K. This implies that A is closed in particular, and therefore A is compact by Lemma 3.39. Since A contains only isolated points, for each $x \in A$ there exists $\varepsilon_x > 0$ so that $N_{\varepsilon_x}(x) \cap K = \{x\}$. The collection $\{N_{\varepsilon_x}\}_{x \in A}$ is an open cover of A which clearly admits no finite subcover, since A is infinite and each neighborhood contains a single point. This contradicts the compactness of A. Therefore A has a limit point in K. $\square$

To prove the converse of Theorem 3.44, we need a preliminary result that gives some uniform control over the size of neighborhoods that fit within a particular covering.

Lemma 3.45 (**Lebesgue covering lemma**) *Suppose that K is a sequentially compact subset of a metric space X. Given an open cover $\{U_\alpha\}_{\alpha \in \mathcal{J}}$, there exists a radius $r > 0$ such that for each $x \subset K$, $N_r(x)$ is contatined in U_α for some $\alpha \in \mathcal{J}$.*

Proof Assume, for the sake of contradiction, that no such r exists. Then for each $k \in \mathbb{N}$ we can find a point $x_k \in K$ for which $N_{1/k}(x_k)$ is not contained in any U_α. Since K is sequentially compact, there exists a subsequence (x_{k_n}) converging to some $y \in K$. Choose $\alpha_0 \in \mathcal{J}$ so that $y \in U_{\alpha_0}$. Since U_{α_0} is open, there exists $\delta > 0$ so that $N_\delta(y) \subset U_{\alpha_0}$. For $k_n > 2/\delta$ we then have

$$N_{1/k_n}(x_{k_n}) \subset N_\delta(y) \subset U_{\alpha_0}$$

by the triangle identity. This contradicts the choice of x_{k_n}. □

Theorem 3.46 *In a metric space, sequential compactness implies compactness.*

Proof Assume that $K \subset X$ is sequentially compact, and let $\{U_\alpha\}$ be an open cover for K. Choose $r > 0$ according to Lemma 3.45, so that for each $x \in K$, $N_r(x) \subset U_\alpha$ for some α.

Fix some $x_1 \in K$, and then choose $x_2 \notin N_r(x_1)$, if possible. We then continue this process as long as possible, choosing, if we can, x_k so that

$$x_k \notin N_r(x_1) \cup \cdots \cup N_r(x_{k-1}).$$

If this process were continued indefinitely, then the result would be an infinite sequence with points separated by distance at least r from each other. Such a sequence cannot have a convergent subsequence, and is thus ruled out by the sequential compactness of K.

Therefore, the process terminates at some finite value of n, for which

$$K \subset N_r(x_1) \cup \cdots \cup N_r(x_n).$$

Since each neighborhood $N_r(x_j)$ lies in some set U_j taken from the cover, we have constructed a finite subcover,
$$K \subset U_1 \cup \cdots \cup U_n.$$
□

The equivalence of these different definitions of compactness means that in metric topology we can choose the formulation that seems best suited to the proof or application at hand. For example, the result of Exercise 3.38 is very easy to prove using sequential compactness, because use of the subspace topology does not affect the convergence of a sequence.

Exercise 3.47 Prove directly that sequentially compact sets are closed and bounded, i.e., give a sequential proof of Theorem 3.37.

Exercise 3.48 Prove directly that closed and bounded subsets of $\mathbb{R}^n$ are sequentially compact, i.e., give a sequential proof of Theorem 3.40.

Exercise 3.49 In a metric space X, suppose that K and F are disjoint sets with K compact and F closed. Prove that the distance

$$d(K, F) := \inf\{d(p, q) : p \in K, q \in F\}$$

is strictly positive.

3.5 Baire Category Theorem

The proof that $\mathbb{R}$ is uncountable that we gave in Theorem 1.14 uses completeness in an essential way, via the monotone sequence theorem. This might suggest that a complete metric space needs be *large* in some topological sense, in order to accommodate all possible limits of Cauchy sequences. However, it not so easy to make this intuition rigorous. After all, a metric space consisting of finitely many points is complete by default, so the number of points is not the relevant measure of size here.

The Baire category theorem gives a precise formulation of this principle. The traditional statement divides subsets of a metric space X into two categories. A set in the *first category* can be written as a countable union of nowhere dense sets. A set is called *nowhere dense* if its closure contains no interior points. In other words, A is nowhere dense if $(\overline{A})^\circ = \emptyset$. These first-category sets are also called *meager*. As this terminology suggests, meager sets are considered small relative to the topology of X.

Example 3.50 Every countable subset of $\mathbb{R}^n$ is meager, because a single point is obviously nowhere dense in the Euclidean topology. However, if we consider $\mathbb{Z}$ as a metric space with the subspace topology inherited from $\mathbb{R}$, then for each $k \in \mathbb{Z}$, the set $\{k\}$ is both open and closed. In this topology, every point of $\mathbb{Z}$ qualifies as an interior point, and hence the only nowhere dense set is the empty set. ◊

Example 3.51 The *Cantor set* C consists of all real numbers in the interval $[0, 1]$ which admit a ternary expansion using only digits 0 and 2. That is, $x \in C$ if we can write

$$x = \sum_{n=1}^{\infty} \frac{a_n}{3^n},$$

where each a_n is 0 or 2. The set of allowable sequences $(a_1, a_2, \dots)$ is uncountable, by Cantor's diagonal argument, and so C is uncountable.

A standard way to construct C is to start from $C_0 = [0, 1]$ and remove successive middle-third intervals. The digit condition $a_1 \neq 1$ corresponds to removing the middle third from $[0, 1]$ leaving $C_1 = [0, \frac{1}{3}] \cup [\frac{2}{3}, 1]$. The condition $a_2 \neq 1$ corresponds to cutting $(\frac{1}{9}, \frac{2}{9})$ and $(\frac{7}{9}, \frac{8}{9})$, and so on. The result is that

$$C = \bigcap_{n=0}^{\infty} C_n,$$

where each C_n is a union of 2^n closed intervals of length 3^{-n}, as illustrated in Fig. 3.3. It follows from this construction that C is closed and contains no open intervals. Therefore C is nowhere dense as a subset of $\mathbb{R}$. ◊

3.5 Baire Category Theorem

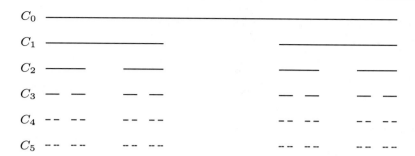

Fig. 3.3 Steps in the Cantor set construction

Baire's *second category* consists of the sets which are not meager, i.e., sets which cannot be written as a countable union of nowhere dense sets. The standard formulation of Baire's theorem says that a complete metric space is of the second category, i.e., not meager. We will first prove an equivalent statement:

Theorem 3.52 *Let X be a complete metric space. Given a countable collection of open dense subsets $U_n \subset X$, the intersection $\cap_{n=1}^{\infty} U_n$ is dense in X.*

Proof Suppose each U_n is open and dense in X and let

$$E := \cap_{n=1}^{\infty} U_n.$$

By Exercise 3.16, to show E is dense in X it suffices to prove that for every non-empty open set $V \in X$, $E \cap V$ is not empty.

Let V be a non-empty open set. Using the density of U_n we can construct a sequence of points x_n and radii r_n as follows. First choose $x_1 \in V \cap U_1$ and $r_1 \leq 1$ such that

$$\overline{N_{r_1}(x_1)} \subset V \cap U_1.$$

Then for each n we choose $x_n \in N_{r_{n-1}}(x_{n-1}) \cap U_n$ and $r_n \leq 1/n$ so that

$$\overline{N_{r_n}(x_n)} \subset N_{r_{n-1}}(x_{n-1}) \cap U_n.$$

By construction, $d(x_n, x_m) \leq 1/n$ for $m > n$, so the sequence is Cauchy and we can take the limit $x := \lim x_n$. The nested construction ensures that $x \in \overline{N_{r_n}(x_n)}$ for all n. It follows that $x \in V \cap U_n$ for all n, and therefore $x \in (\cap_{n=1}^{\infty} U_n) \cap V$. □

The corresponding result for closed sets is obtained by taking complements.

Exercise 3.53 Show that a closed set has empty interior if and only if its complement is dense.

If X is written as a union of closed sets $\cup F_n$, then each F_n^c is open and $\cap F_n^c = \emptyset$. Then Theorem 3.52 implies that F_n^c is not dense for some n. Combining this observation with Exercise 3.53 yields the following:

Corollary 3.54 *Let X be a complete metric space. If $X = \cup_{n=1}^{\infty} F_n$ where each F_n is closed, then at least one F_n has a nonempty interior.*

Since the closure of a nowhere dense set has empty interior, the corollary implies in particular that complete metric spaces are not meager. Thus Corollary 3.54 is a statement of the Baire category theorem.

The Baire category theorem leads to quick proofs of some rather deep results. For example, we can rule out completeness for certain spaces:

Example 3.55 We claim that there is no norm under which the vector space $V := \{(a_1, a_2, \ldots) : a_n \in \mathbb{C}\}$ is complete. Let $\|\cdot\|$ be a norm on V. The subspaces

$$V_n := \{(a_1, \ldots, a_n, 0, \ldots)\},$$

are finite-dimensional normed vector spaces. By Corollary 3.28, each V_n is complete as a subspace and therefore closed as a subset. Moreover, V_n has no interior points. Therefore, each V_n is nowhere dense and $V = \cup_n V_n$ is meager. By the Baire theorem, V is not complete. ◊

Functions on Metric Spaces 4

In this chapter we will discuss maps from one metric space to another. The special case of real- or complex-valued functions is of special importance and will be developed further in the next chapter. For now we focus on the most general definitions and properties. We start by introducing function limits and continuity, and then consider limits of sequences of functions.

4.1 Continuity

Let (X, d_X) and (Y, d_Y) be metric spaces. On a domain $E \subset X$, consider a function $f : E \to Y$. We say that f is *continuous* at $x_0 \in E$ if for every $\varepsilon > 0$ there exists $\delta > 0$ such that

$$d_X(x, x_0) < \delta \implies d_Y(f(x), f(x_0)) < \varepsilon \qquad (4.1)$$

for $x \in E$. This condition is trivially satisfied if x_0 is isolated, because in this case we can choose ε small enough that $N_\varepsilon(x_0)$ contains no point of E other than x_0. If continuity holds at every point of the domain, then f is said to be continuous "on E."

Example 4.1 On $\mathbb{R}$ consider the monomial function $f(x) = x^n$ with $n \in \mathbb{N}$. (This proof of continuity should be familiar from calculus.) By the polynomial identity

$$x^n - x_0^n = (x - x_0)(x^{n-1} + x^{n-2}x_0 + \cdots + x_0^{n-1}), \qquad (4.2)$$

we can estimate for $|x - x_0| \le 1$,

$$\left|x^{n-1} + x^{n-2}x_0 + \ldots x_0^{n-1}\right| \leq C_{x_0},$$

where
$$C_{x_0} = n(|x_0| + 1)^{n-1}.$$

This gives the estimate
$$|f(x) - f(x_0)| \leq C_{x_0}|x - x_0|,$$

which establishes continuity at x_0. ◇

Example 4.2 Consider a linear map $T : \mathbb{R}^n \to \mathbb{R}^m$, which we can write in terms of an $m \times n$ matrix A:
$$Tv = A \cdot v.$$

In terms of the max norm
$$\|Tv\|_{\max} = \max_i \left|\sum_{j=1}^n A_{ij} v_j\right|$$
$$\leq C\|v\|_{\max},$$

where $C = n \max_{ij} |A_{ij}|$. This proves that T is continuous with respect to the max norm, since by linearity
$$\|Tv - Tw\|_{\max} \leq C\|v - w\|_{\max}.$$
◇

We saw that all finite-dimensional norms are equivalent in Theorem 3.27. Therefore, Example 4.2 shows that all linear maps on finite dimensional normed vector spaces are continuous. This is not true for infinite dimensional vector spaces.

Example 4.3 Recall the spaces of sequences ℓ^p defined in Example 3.4 and Exercise 3.5. Let W denote the space ℓ^1, equipped with the ℓ^∞ norm
$$\|\alpha\|_\infty = \sup_j |\alpha_j|.$$

Thus W and ℓ^1 denote the same set, but with different norms.

Let $S : W \to \ell^1$ denote the identity map $\alpha \mapsto \alpha$. Despite being a very simple linear map, S is not continuous. For example, for $n \in \mathbb{N}$ consider the sequence defined by
$$\alpha = (\underbrace{1, \ldots, 1}_{n}, 0, \ldots).$$

Then $\alpha \in W$ with $\|\alpha\|_\infty = 1$. Since
$$\|S\alpha\|_1 = \|\alpha\|_1 = n,$$

4.1 Continuity

the norm of $S\alpha$ can be arbitrarily large relative to the norm of α.

On the other hand, the identity map in the direction $\ell^1 \to W$ is continuous, because $\|\alpha\|_\infty \leq \|\alpha\|_1$ for all α. $\diamond$

Exercise 4.4 Let (X, d) be a metric space. Given a non-empty subset $A \subset X$ the distance from a point x to A is defined as

$$d(A, x) := \inf\{d(y, x) : y \in A\}$$

Show that $d(A, \cdot)$ is continuous as a function $X \to \mathbb{R}$.

Exercise 4.5 Suppose that $f : X \to Y$ is continuous. For a subset $E \subset X$, prove that

$$f(\overline{E}) \subset \overline{f(E)}.$$

The definition of continuity is closely related to the concept of a function limit. For $f : E \to Y$ as above, for $y \in Y$ and x_0 a limit point of E, we say that $f(x) \to y$ as $x \to x_0$, or

$$\lim_{x \to x_0} f(x) = y,$$

if for every $\varepsilon > 0$ there is a $\delta > 0$ so that

$$d_X(x, x_0) < \delta \implies d_Y(f(x), x_0) < \varepsilon$$

for $x \in E$. Function limits can be equivalently formulated in terms of convergence of sequences, as in the following:

Exercise 4.6 For a function $f : E \to Y$ and x_0 a limit point of E, show that $f(x) \to y$ as $x \to x_0$ if and only if $f(x_n) \to y$ for every sequence $(x_n) \in E$ with $x_n \to x_0$.

Note that in the function limit definition p is assumed to be a limit point of the domain and it is irrelevant whether or not f is defined at p. Continuity is only defined at points in the domain. However, we noted that continuity holds trivially at an isolated point. We can thus clarify the relationship between continuity and function limits as follows:

Lemma 4.7 *The function $f : E \to Y$ is continuous at $x_0 \in E$ if and only if one of the following conditions holds:*

(i) x_0 is a limit point of E and $\lim_{x \to x_0} f(x) = f(x_0)$;
(ii) x_0 is an isolated point of E.

The definition of continuity given at the start of this section is probably familiar to the reader from standard definition for real functions. The following result gives an equivalent topological formulation, which refers only to open sets and does not mention the distance function.

Theorem 4.8 *A map $f : E \to Y$ is continuous if and only if the preimage of an open set in Y is open relative to E.*

Proof Assume that f is continuous on E, and let $U \subset Y$ be open. Note that there is nothing to prove if $f^{-1}(U)$ is empty. Suppose there is a point $x \in E$ such that $f(x) \in U$. Since U is open, we can choose $\varepsilon > 0$ so that $N_\varepsilon(f(x)) \subset U$. Condition (4.1) then gives $\delta > 0$ such that
$$f(N_\delta(x) \cap E) \subset N_\varepsilon(f(x)), \tag{4.3}$$
implying that $N_\delta(x) \cap E \subset f^{-1}(U)$. This works for all $x \in f^{-1}(U)$, proving that $f^{-1}(U)$ is open relative to E.

The converse argument is similar. Suppose that the preimage of an open set is open. For $x \in E$ and $\varepsilon > 0$ this implies that $f^{-1}(N_\varepsilon(f(x)))$ is open relative to E. Therefore, x has a neighborhood inside $f^{-1}(N_\varepsilon(f(x)))$. In other words, there exists $\delta > 0$ such that
$$N_\delta(x) \cap E \subset f^{-1}(N_\varepsilon(f(x))).$$
This is equivalent to the continuity condition (4.3). □

The topological interpretation of continuity expressed in Theorem 4.8 is convenient even in the metric case, because it often leads to cleaner arguments. For example, since preimages are easy to track through compositions of functions, we obtain a very short proof of the composition rule:

Theorem 4.9 *Suppose $f : X \to Y$ and $g : Y \to Z$ are continuous functions. The composition $g \circ f$ is a continuous map $X \to Z$.*

Proof This result follows immediately from Theorem 4.8, since
$$(g \circ f)^{-1}(U) = f^{-1}(g^{-1}(U)).$$
□

Since open sets are the complements of closed sets, and taking preimages preserves the complement relationship, Theorem 4.8 implies the corresponding statement for closed sets:

4.1 Continuity

Corollary 4.10 *A map $f : E \to Y$ is continuous if and only if the preimage of a closed set in Y is closed relative to E.*

It is important to remember that the continuity implies nothing about the image of an open or closed set. For example, a constant function is continuous by default, but the image of any open set is a single point. A function that takes open sets to open sets is called an *open map*.

The images and preimages of a compact set under a continuous maps behave differently from open or closed sets. The preimage of a compact set need not be compact. For example, if $f : X \to Y$ and Y is compact, then $f^{-1}(Y) = X$, whether or not X is compact. On the other hand, the following result shows that continuous maps preserve compactness.

Exercise 4.11 Prove that the image of a compact set under a continuous function is compact. Give two different proofs, first with the open cover definition and then using sequential compactness. $\Diamond$

When applied to real-valued functions, Exercise 4.11 shows that a continuous function on a compact domain achieves both a minimum and maximum value.

Theorem 4.12 (extreme value theorem) *If $f : X \to \mathbb{R}$ is continuous and $K \subset X$ is compact, then there exist points $q_\pm \in K$ such that*

$$f(q_-) \leq f(x) \leq f(q_+)$$

for all $x \in K$.

Proof The hypotheses imply that $f(K)$ is a compact subset of $\mathbb{R}$ by Exercise 4.11. Exercise 1.7 implies that there exists a sequence in $f(K)$ converging to $\sup f(K)$. Since $f(K)$ is closed and bounded, this implies that $\sup f(K) \in f(K)$. Hence there exists $q_+ \in K$ such that

$$f(q_+) = \sup f(K).$$

The same reasoning applies to $\inf f(K)$. $\square$

Another useful implication of Exercise 4.11 is a guarantee of continuity for inverse functions, provided the domain is compact.

Theorem 4.13 *If K is compact and $f : K \to Y$ is bijective and continuous, then f^{-1} is continuous.*

Proof Bijectivity guarantees that the inverse function $g = f^{-1}$ is defined. Suppose $E \subset K$ is closed. Then E is compact by Lemma 3.39, and hence $f(E)$ is compact by Exercise 4.11.

Since $f(E) = g^{-1}(E)$, this proves that the preimage of a closed set under g is closed. Hence g is continuous by Corollary 4.10. $\square$

The result of Theorem 4.13 can fail without the compactness assumption. For example, the function $f(\theta) = e^{i\theta}$ defines a continuous and bijective map from $[0, 2\pi)$ to the unit circle in $\mathbb{C}$, but f^{-1} is discontinuous at 1.

To conclude this section we consider the mapping properties of connected sets under continuous functions. (Connected sets were defined in Sect. 3.2.1.)

Exercise 4.14 Prove that the image of a connected set under a continuous function is connected.

4.1.1 Uniform Continuity

Suppose f is continuous on a domain E. Given $x_0 \in E$ and $\varepsilon > 0$ we can choose $\delta > 0$ so that the continuity condition (4.1). This does not mean that the same choice of δ works for every point, however. So we should really write δ as δ_{x_0} if we are making this choice at multiple points.

Some applications, particularly in integration, require a stronger notion of continuity that eliminates the dependence on the point. A function $f : E \to Y$ is *uniformly continuous* if for every $\varepsilon > 0$ there exists $\delta > 0$ such that

$$d(p, q) < \delta \implies d(f(p), f(q)) < \varepsilon \qquad (4.4)$$

for all points $p, q \in E$.

Example 4.15 Consider again the real function $f(x) = x^n$ as in Example 4.1. The ratio between ε and δ at y was given by

$$C_y = n(|y| + 1)^{n-1}.$$

For $n = 1$ we have $C_y = 1$ so this is uniform on all of $\mathbb{R}$. For $n > 1$, it is easy to check that the function will not be uniformly continuous on all of $\mathbb{R}$. For example, if $x > 0$ and $t > 0$, then

$$(x + t)^n - x_n \geq nx^{n-1}t.$$

This difference becomes arbitrarily large as $x \to \infty$, for any fixed t.

On the other hand, if we restrict the domain to a compact interval K then we can take C to be the maximum of C_y over K to conclude that

$$|f(x) - f(y)| \leq C|x - y|$$

for $x, y \in K$. Thus f is uniformly continuous on any compact interval. $\diamond$

Example 4.16 Suppose $f(z)$ is a complex function defined by the power series

$$f(z) := \sum_{k=0}^{\infty} c_k z^k,$$

with radius of convergence $R > 0$. We claim that f is uniformly continuous on the disk $D_r := \{|z| \leq r\}$ for $r < R$.

For $z, w \in D_r$, the absolute convergence of the power series allows us to combine terms

$$f(z) - f(w) = \sum_{k=1}^{\infty} c_k (z^k - w^k).$$

We can then extract a factor of $(z - w)$ from each term,

$$f(z) - f(w) = (z - w) \sum_{k=1}^{\infty} c_k \left(z^{k-1} + z^{k-2} w + \cdots + w^{k-1} \right),$$

and estimate

$$|f(z) - f(w)| \leq |z - w| \sum_{k=1}^{\infty} k |c_k| r^{k-1}. \tag{4.5}$$

Since $k^{1/k} \to 1$ by Example 1.4, the result of Exercise 1.20 implies that

$$\limsup (k|c_k|)^{1/k} = \limsup |c_k|^{1/k}. \tag{4.6}$$

Hence, by Theorem 2.16 the series in (4.5) is convergent for $r < R$ and we obtain

$$|f(z) - f(w)| \leq C_r |z - w| \tag{4.7}$$

for $z, w \in D_r$. This proves uniform continuity on D_r for any $r < R$. ◇

Examples 4.15 and 4.16 are illustrative of a general phenomenon, namely that the restriction of a continuous function to a compact set is necessarily uniformly continuous.

Exercise 4.17 Suppose that $f : (0, 1) \to \mathbb{R}$ is uniformly continuous. Prove that f admits an extension to a continuous function on the interval $[0, 1]$.

In (4.7) we established uniform continuity by proving something stronger, a qualitative estimate on the distance between image points. We have actually seen the same notion in many previous examples. A function $f : E \to Y$ is *Lipschitz continuous* if there exists a constant $c > 0$ such that

$$d(f(x), f(y)) \leq c d(f(x), f(y)) \tag{4.8}$$

for all $x, y \in E$.

Example 4.18 To illustrate the difference between uniform and Lipschitz continuity, consider the function $f(x) = \sqrt{x}$ on $[0, 1]$. We can easily verify that f is continuous on $[0, 1]$, and therefore uniformly continuous by Exercise 4.17. At 0 it is straightforward to check that $\sqrt{x} \to 0$ as $x \to 0$, and at a point $y > 0$ we can use the identity

$$\sqrt{x} - \sqrt{y} = \frac{x - y}{\sqrt{x} + \sqrt{y}}. \tag{4.9}$$

However, (4.9) also shows that f is not Lipschitz since the denominator on the right side approaches zero at the origin. ◊

A Lipschitz map for which the constant in (4.8) satisfies $c < 1$ is called a *contraction*. Many applications of contractions are based on the following result, sometimes called the *Banach fixed-point theorem*.

Exercise 4.19 Let (X, d) be a complete metric space and suppose that $f : X \to X$ is a contraction. For any $x_1 \in X$, prove that the sequence defined iteratively by $x_k = f(x_{k-1})$ converges to a unique point y such that $f(y) = y$.

4.2 Sequences of Functions

Consider a sequence of functions $f_n : X \to Y$, where X and Y are metric spaces. We say that (f_n) converges *pointwise* to a function f if

$$\lim_{n \to \infty} f_n(x) = f(x)$$

for each $x \in X$. This is the default definition of convergence for function sequences, but it is too weak for many applications. Sometimes we need to insist on a uniform rate of convergence at all points. We say that $f_n \to f$ *uniformly* on a subset $A \subset X$ if

$$\lim_{n \to \infty} \left(\sup_{x \in A} d_Y(f_n(x), f(x)) \right) = 0. \tag{4.10}$$

In other words, for every $\varepsilon > 0$ there exists $N \in \mathbb{N}$ so that $n \geq N$ implies that

$$d_Y(f_n(x), f(x)) < \varepsilon$$

for all $x \in A$.

Example 4.20 Consider the sequence $f_n(x) = x^n$ on the interval $[0, 1)$, as shown in Fig. 4.1. Clearly $f_n \to 0$ pointwise as $n \to \infty$. However, since

4.2 Sequences of Functions

$$\sup_{x\in[0,1)} |x^n| = 1$$

for all n, the convergence fails to be uniform. The choice of domain makes all the difference here. Convergence is uniform on $[0, b]$ for any $b < 1$, because

$$\sup_{x\in(0,b]} |x^n| = b^n$$

Example 4.21 Consider a power series ◇

$$f(z) := \sum_{k=0}^{\infty} c_k z^k,$$

with radius of convergence $R > 0$, as introduced in Sect. 2.4. We claim that the convergence is uniform on the closed disk $\{|z| \leq r\}$ for $r < R$. To see this pick $r < r_1 < R$. The definition of R implies that

$$|c_k z^k| \leq \left(\frac{|z|}{r_1}\right)^k$$

for all but finitely many k. For $|z| \leq r$ this gives a uniform estimate

$$\sum_{k=n}^{\infty} |c_k z^k| \leq \sum_{k=n}^{\infty} \left(\frac{r}{r_1}\right)^k,$$

which shows that the series converges uniformly on $\{|z| \leq r\}$. ◇

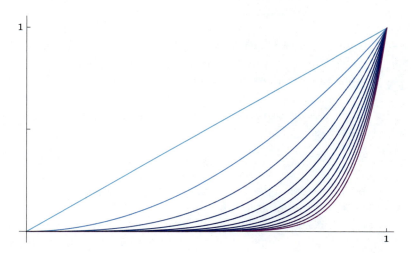

Fig. 4.1 On $(0, 1)$, the functions x^n converge to 0 pointwise as $n \to \infty$ but not uniformly

Continuity of functions is not necessarily preserved under pointwise convergence. For instance, the functions f_n from Example 4.20 are continuous when extended to the domain $[0, 1]$, but the pointwise limit is discontinuous at $x = 1$. Not coincidentally, this is the point where convergence fails to be uniform. The following result shows that uniform convergence does indeed preserve continuity.

Exercise 4.22 Suppose each function $f_n : E \to Y$ is continuous at some $q \in E$. If $f_n \to f$ uniformly on E, then f is continuous at q. ◊

On a metric space X, we denote by $C(X; \mathbb{C})$ the space of continuous complex-valued functions. If X is compact then each positive continuous function has a positive maximum value by Theorem 4.12. Thus we can define a metric based on the norm

$$\|f\|_{\max} := \max_X |f|. \tag{4.11}$$

To say that $f_k \to f$ in $C(X; \mathbb{C})$ means precisely that the sequence converges uniformly on X.

Exercise 4.23 Assuming X is compact, prove that $C(X; \mathbb{C})$ is complete under the metric defined by the norm (4.11).

For real-valued continuous functions on a compact metric space, uniform convergence can be guaranteed if oscillations are avoided. A sequence of real-valued functions f_n is *monotone* if the numerical sequence $(f_n(x))_{n=1}^\infty$ is monotone for each x. Note that for the following result we need an extra assumption that the limit function is continuous, since this is not guaranteed by the other hypotheses.

Theorem 4.24 (**Dini**) *Suppose (X, d) is a compact measure space and let (f_n) be a sequence in $C(X, \mathbb{R})$ converging pointwise to $f \in C(X, \mathbb{R})$. If (f_n) is monotone then $f_n \to f$ in $C(X, \mathbb{R})$, i.e., the convergence is uniform.*

Proof We can assume that the sequence is increasing by switching f_n to $-f_n$ if needed. Given $\varepsilon > 0$, define the sets

$$K_n := \{x \in K : f(x) - f_n(x) \geq \varepsilon\}.$$

Since K_n is the preimage of the closed set $[\varepsilon, \infty)$ under the continuous function $f - f_n$, K_n is closed. Thus K_n is compact since X is compact.

The fact that (f_n) is increasing means the sets are nested $K_n \supset K_{n+1}$. Furthermore, the pointwise convergence $f_n \to f$ implies that $\cap K_n = \emptyset$. By Exercise 3.42 one of the sets, say K_N must be empty. This means that

$$\|f - f_n\|_{\max} < \varepsilon \quad \text{for all } n \geq N.$$

□

4.3 Stone-Weierstrass Theorem

In 1885 Karl Weierstrass proved that a continuous function on a compact interval in $\mathbb{R}$ admits a uniform approximated by polynomials. This result admits a broad generalization in the context of compact metric spaces, proven by Marshall Stone in 1937.

For a compact metric space X, let $C(X; \mathbb{F})$ denote the space of complex-valued functions $X \to \mathbb{F}$, where $\mathbb{F}$ stands for the field $\mathbb{R}$ or $\mathbb{C}$. We give this space the topology of uniform convergence as defined by the max norm (4.11). The space $C(X; \mathbb{F})$ also has the structure of an algebra over $\mathbb{F}$, meaning that it is closed under the operations of pointwise addition and multiplication,

$$(f + g)(x) := f(x) + g(x), \quad (fg)(x) := f(x)g(x),$$

as well as scalar multiplication. We first consider the real case.

Theorem 4.25 (**Stone-Weierstrass**) *For a compact metric space X, let $\mathcal{A}$ be a subalgebra of $C(X; \mathbb{R})$ such that*

(i) $\mathcal{A}$ contains the constant functions;
(ii) $\mathcal{A}$ separates points, meaning given $p \neq q \in K$, there exists $f \in \mathcal{A}$ with $f(p) \neq f(q)$.

Then $\mathcal{A}$ is dense in $C(X; \mathbb{R})$.

The classical Weierstrass theorem is the special case with $X = [a, b] \subset \mathbb{R}$ and $\mathcal{A}$ the algebra of real polynomials on $[a, b]$. This clearly satisfies the hypotheses, and so we immediately obtain the folllowing:

Corollary 4.26 *Let $f : [a, b] \to \mathbb{R}$ be continuous. Given $\varepsilon > 0$, there exists a polynomial $p(x)$ such that*

$$|f(x) - p(x)| \leq \varepsilon$$

for all $x \in [a, b]$.

We will present an elementary proof of Theorem 4.25 adapted from [Brosowski-Deutsch 1981]. The first step is to show that the algebra $\mathcal{A}$ allows us approximate the indicator function of a set.

Lemma 4.27 *Suppose $\mathcal{A} \subset C(X; \mathbb{R})$ satisfies the hypotheses of Theorem 4.25. Given a closed set $B \subset X$ and a point $p \notin B$, there exists a open set $V_p \ni p$, disjoint from B, with the following property: Given $\varepsilon > 0$ there exists $\psi \in \mathcal{A}$ with $0 \leq \psi \leq 1$ and satisfying*

$$\psi|_{V_p} < \varepsilon, \qquad \psi|_B > 1 - \varepsilon.$$

Proof By the assumptions on $\mathcal{A}$, for each $x \in B$ we can find a function that vanishes at p but not at x. Squaring this function gives $h_x \in \mathcal{A}$ such that $h_x(p) = 0$ and $h_x(x) > 0$. And after multiplying by a constant, if necessary, we can assume $0 \leq h_x \leq 1$.

Because B is compact and the sets $\{h_x > 0\}$ are open, there exist a finite number of points $x_1, \ldots, x_m$ such that

$$B \subset \bigcup_{j=1}^{m} \{h_{x_j} > 0\}.$$

The function $h \in \mathcal{A}$ defined as the average

$$h := \frac{1}{m} \sum_{j=1}^{m} h_{p_j}$$

then satisfies $0 \leq h \leq 1$, $h(p) = 0$ and $h > 0$ on B.

To define the neighborhood V_p, by the extreme value theorem (Theorem 4.12) we can choose $c > 0$ so that

$$h|_B > c.$$

We then set

$$V_p := \{h < c/2\}.$$

To complete the proof, the plan is to use h to construct the function ψ for a given ε.

For $t \in [0, 1]$ and $n \in \mathbb{N}$ we have the basic inequalities

$$1 - nt \leq (1 - t)^n \leq \frac{1}{1 + nt}. \tag{4.12}$$

The left-hand side is the classic *Bernoulli inequality*, which is easily proven by induction on n. The right-hand inequality follows from Bernoulli by the observation that $(1 - t)(1 + t) \leq 1$. (Of course, these are also easily proven by calculus methods.)

For $k, n \in \mathbb{N}$, if we set

$$\psi := 1 - (1 - h^k)^n,$$

then (4.12) and the bounds on h we have

$$\psi|_{V_p} < \frac{nc^k}{2^k}, \qquad \psi|_B > 1 - \frac{1}{nc^k}.$$

If we then set $n = l^k$ for $l \in \mathbb{N}$ then the estimates reduce to

4.3 Stone-Weierstrass Theorem

$$\psi|_{V_p} < a^k, \qquad \psi|_B > 1 - b^k,$$

where $a = lc/2$ and $b = 1/lc$. Since $c \in (0, 1)$, we can choose an integer $l \in (1/c, 2/c)$, which gives $a < 1$ and $b < 1$. The desired ψ is then obtained by taking k sufficiently large. □

Figure 4.2 illustrates the functions defined in Lemma 4.27 in the polynomial case. The tricky issue in this construction is the need to fix a neighborhood that is independent of ε. The uniformity provided by Lemma 4.27 allows us to improve the result to the following:

Corollary 4.28 *Suppose A and B are disjoint closed subsets of X. For $0 < \varepsilon < 1$ there exists $\chi \in \mathcal{A}$ such that $0 \le \chi \le 1$ and*

$$\chi|_A < \varepsilon, \qquad \chi|_B > 1 - \varepsilon.$$

Proof Since A is compact, we can cover it cover it with a finite number of neighborhoods of the type constructed in Lemma 4.27,

$$A \subset V_1 \cup \cdots \cup V_m.$$

For $\varepsilon > 0$ we can then construct functions $\psi_1, \ldots, \psi_m \in \mathcal{A}$ such that $0 \le \psi_j \le 1$ and

$$\psi_j|_{V_j} < \frac{\varepsilon}{m}, \qquad \psi_j|_B > 1 - \frac{\varepsilon}{m}.$$

The function $\chi := \prod_{j=1}^m \psi_j$ then has the stated properties. □

Corollary 4.28 allows us to construct approximate step functions using elements of $\mathcal{A}$. With these approximations available, it is relatively straightforward to establish the uniform approximation of continuous functions.

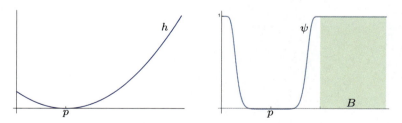

Fig. 4.2 In the polynomial case we could take $h(x) = (x - p)^2$ in Lemma 4.27. The plot on the right shows the function ψ associated to the subset B

Proof of Theorem 4.25 Suppose that $\mathcal{A} \subset C(X;\mathbb{R})$ satisfies the two hypotheses. Given $f \in C(X;\mathbb{R})$, our goal is to approximate f uniformly from within $\mathcal{A}$. Because $\mathcal{A}$ contains constants and f is bounded, we can assume that $0 \leq f \leq 1$.

For $n \in \mathbb{N}$ and $j = 0, \ldots, n$, define the disjoint pairs of closed sets

$$A_j := \left\{ f \leq \frac{j}{n} \right\}, \qquad B_j := \left\{ f \geq \frac{j+1}{n} \right\}.$$

For each j we apply Corollary 4.28 to define $\chi_j \in \mathcal{A}$ satisfying $0 \leq \chi_j \leq 1$ and

$$\chi_j|_{A_j} < \frac{1}{n}, \qquad \chi_j|_{B_j} > 1 - \frac{1}{n}. \tag{4.13}$$

Our proposed approximation to f is

$$g_n := \frac{1}{n} \sum_{j=0}^{n-1} \chi_j.$$

Given a point $x \in X$ there exists a unique $k \in \{0, \ldots, n-1\}$ so that

$$\frac{k}{n} < f(x) \leq \frac{k+1}{n}. \tag{4.14}$$

This means that $x \in A_j$ for $j \geq k$, which gives an estimate

$$g_n(x) \leq \frac{1}{n} \sum_{j=0}^{k-1} 1 + \frac{1}{n} \sum_{j=k}^{n-1} \frac{1}{n}$$

$$\leq \frac{k+1}{n}$$

$$< f(x) + \frac{1}{n}.$$

On the other hand, (4.14) implies that $x \in B_j$ for $j \leq k-1$, which gives

$$g_n(x) \geq \frac{1}{n} \sum_{j=0}^{k-1} \left(1 - \frac{1}{n}\right)$$

$$\geq \frac{k-1}{n}$$

$$\geq f(x) - \frac{2}{n}.$$

Since these estimates apply to all $x \in X$, we have shown that

$$\|f - g_n\|_{\max} \leq \frac{2}{n}.$$

4.3 Stone-Weierstrass Theorem

Therefore $g_n \to f$ in $C(X; \mathbb{R})$. □

With the real case established, the complex version is a relatively simple extension.

Corollary 4.29 *Let X be a compact metric space, and suppose $\mathcal{B}$ is a subalgebra of $C(X; \mathbb{C})$ which is closed under complex conjugation. If $\mathcal{B}$ contains the constant functions and separates points, then $\mathcal{B}$ is dense in $C(X; \mathbb{C})$.*

Proof Let $\mathcal{B}_\mathbb{R}$ denote the set of real-valued functions in $\mathcal{B}$. Since $\mathcal{B}$ contains constants, it follows that $\mathcal{B}_\mathbb{R}$ contains the real-valued constant functions. By hypothesis, for $p \neq q$ in X there exists $f \in \mathcal{B}$ such that $f(p) \neq f(q)$. After shifting by a constant and rescaling, we can assume that $f(p) = 1$ and $f(q) = 0$. Then, because $\mathcal{B}$ is closed under conjugation, the function $u := f + \overline{f}$ lies in $\mathcal{B}_\mathbb{R}$. Since $u(p) \neq u(q)$, this proves that $\mathcal{B}_\mathbb{R}$ separates points. Therefore $\mathcal{B}_\mathbb{R}$ is dense in $C(X; \mathbb{R})$ by Theorem 4.25. A function in $C(X; \mathbb{C})$ can therefore be approximated uniformly from $\mathcal{B}$ by approximating its real and imaginary parts by functions in $\mathcal{B}_\mathbb{R}$. □

Example 4.30 For $X = [0, \pi]$, let $\mathcal{B}$ be the algebra of complex trigonometric polynomials, i.e., linear combinations of the complex exponentials $f_k(x) := e^{ikx}$ for $k \in \mathbb{Z}$. The constant functions are included as multiples of f_0, and $\mathcal{B}$ is clearly closed under complex conjugation. Since $\mathcal{B}$ contains the function $\cos x = \frac{1}{2}(f_1(x) - f_{-1}(x))$, which is injective on $[0, \pi]$, the algebra separates points. Therefore, the Stone-Weierstrass theorem implies that functions in $C[0, \pi]$ can be uniformly approximated by trigonometric polynomials. ◊

Real Functions 5

In this final chapter our goal is to illustrate the tools developed in previous sections in the most basic context, namely real-valued functions defined on intervals. We will review some of the fundamentals of calculus, but the main point here is to demonstrate some concrete applications of the abstract concepts of metric topology.

5.1 Limits and Continuity

Let $I \subset \mathbb{R}$ be an interval as defined in Sect. 1.1. For a function $f : I \to \mathbb{R}$ and x_0 a point in the closure $\overline{I}$, the function limit

$$\lim_{x \to x_0} f(x) = y, \qquad (5.1)$$

means that for every $\varepsilon > 0$, there exists $\delta > 0$ so that $|f(x) - y| < \varepsilon$ for all $x \in E$ with $|x - x_0| < \delta$.

The algebraic properties of sequence limits derived in Exercises 1.6 and 1.8 carry over directly to real function limits, with essentially the same proofs.

Lemma 5.1 *Let f and g be real functions defined on an interval $I \subset \mathbb{R}$. For $x_0 \in \overline{I}$ suppose that x_0 is a limit point of E and that $f(x) \to a$ and $g(x) \to b$ as $x \to x_0$. Then*

$$\lim_{x \to x_0} [f(x) + g(x)] = a + b, \qquad \lim_{x \to x_0} f(x)g(x) = ab.$$

Furthermore, If $f(x) \leq g(x)$ for all $x \in I$ then $a \leq b$.

Real function limits are extended to cases where either x_0 or y is $\pm\infty$ by using the topology of $(\mathbb{R}_\infty, d_\infty)$ introduced in Example 3.2. This interpretation agrees with the definition of

(5.1) when x_0 and y are both finite, and also covers all of the possible infinite cases. Since the distance function d_∞ is cumbersome to use in practice, so we normally do not refer to this directly. Instead, we can formulate the infinite limits in terms of neighborhoods. In the topology of $(\mathbb{R}_\infty, d_\infty)$ a neighborhood of $+\infty$ has the form $(m, \infty]$ for $m \in \mathbb{R}$, and a neighborhood of $-\infty$ can be written as $[-\infty, m)$ of the infinite cases. The statement $f(x) \to \infty$ as $x \to x_0 \subset \mathbb{R}$ therefore means that for each $m > 0$, there exists $\delta > 0$ so that for $x \in E$,
$$f(x) > m \quad \text{if } |x - x_0| < \delta.$$
Similarly for $y \in \mathbb{R}$ and I unbounded above, the statement $f(x) \to y$ as $x \to \infty$ means that for every $\varepsilon > 0$ there exists $m > 0$ so that
$$|f(x) - y| < \varepsilon \quad \text{if } x > m.$$
The algebraic results of Lemma 5.1 obviously require the limits a and b to be finite, but the ordering statement still applies to limits of $\pm\infty$.

As in the sequence case, the crucial limitation on the use of Lemma 5.1 is the hypothesis that the limits exist. We can work around this using upper and lower function limits. For $f : I \to \mathbb{R}$ and real $x_0 \in \overline{I}$, define
$$\limsup_{x \to x_0} f(x) := \inf_{\delta > 0} \left[\sup\{f(x) : x \in E, |x - x_0| < \delta\} \right]$$
and
$$\liminf_{x \to x_0} f(x) := \sup_{\delta > 0} \left[\inf\{f(x) : x \in E, |x - x_0| < \delta\} \right].$$
The infinite cases are treated similarly. For example, if I is unbounded above then
$$\limsup_{x \to \infty} f(x) := \inf_{m > 0} \left[\sup\{f(y) : y \in E, x > m\} \right].$$
By the supremum property of $\mathbb{R}$, these upper and lower limits exist (as values in $\mathbb{R}_\infty$) under any circumstances.

The main useful properties of upper and lower function limits are summarized in the following pair of lemmas, whose proofs are very similar to the sequential cases:

Lemma 5.2 *For real-valued functions defined on an interval $I \subset \mathbb{R}$ and α a limit point of I in $\mathbb{R}_\infty$:*

(i) *$f(x)$ has limit as $x \to \alpha$ if and only if*
$$\liminf_{x \to \alpha} f(x) = \limsup_{x \to \alpha} f(x).$$

(ii) *If $f \leq g$ on I then*

5.1 Limits and Continuity

$$\liminf_{x \to \alpha} f(x) \leq \liminf_{x \to \alpha} g(x), \quad \limsup_{x \to \alpha} f(x) \leq \limsup_{x \to \alpha} g(x).$$

(iii) If f and g are bounded then

$$\limsup_{x \to \alpha}[f(x) + g(x)] \leq \limsup_{x \to \alpha} f(x) + \limsup_{x \to \alpha} g(x)$$

and

$$\liminf_{x \to \alpha}[f(x) + g(x)] \geq \liminf_{x \to \alpha} f(x) + \liminf_{x \to \alpha} g(x)$$

(iv) If $f(x) \to c \in \mathbb{R}$ as $x \to \alpha$ and g is bounded then

$$\limsup_{x \to \alpha} f(x)g(x) = c \limsup_{x \to \alpha} g(x), \quad \liminf_{x \to \alpha} f(x)g(x) = c \limsup_{x \to \alpha} g(x)$$

A common strategy for proving that $f(x) \to y$ as $x \to \alpha$ is to show that

$$\limsup_{x \to \alpha} |f(x) - y| = 0. \tag{5.2}$$

This proves that the limit exists, since we automatically have $\liminf |\ldots| \geq 0$. Lemma 5.2 provides the essential tools to estimate the left side of (5.2).

5.1.1 Asymptotics and Order Notation

In this section we will introduce some helpful notations used to describe the limiting behavior of a function. These conventions first appeared in analytic number theory in the early 20th century, and have since become widely used in analysis and many other fields.

The first *order notation*, generally referred to as "big O," is to write

$$f(x) = O(\phi(x)) \quad \text{as } x \to \alpha \tag{5.3}$$

to signify that $f(x)/\phi(x)$ is bounded as $x \to \alpha$. For $\alpha \in \mathbb{R}$ this means that there exists $C, r > 0$ such that

$$|f(x)| \leq C|\phi(x)| \quad \text{for } 0 < |x - \alpha| < r.$$

The infinite cases are interpreted using neighborhoods in $\mathbb{R}_\infty$. For example $f(x) = O(\phi(x))$ as $x \to \infty$ means there exist $C, m > 0$ such that

$$|f(x)| \leq C|\phi(x)| \quad \text{for } x > m.$$

Exercise 5.3 For the polynomial $p(x) = a_n x^n + \cdots + a_1 x$, where $n > 1$, show that $p(x) = O(x^n)$ as $x \to \infty$ and $p(x) = O(x)$ as $x \to 0$.

The use of an equals sign in (5.3) might seem inconsistent, since $O(\phi)$ describes an estimate rather than an exact limit. The convention here is to interpret $O(\phi)$ as an unspecified function, so that it can be manipulated like other functions. This makes the notation particularly useful for describing error terms. For example, from the power series expansion of $\sin(x)$ we can deduce that

$$\sin x = x + O(x^3) \tag{5.4}$$

as $x \to 0$.

Exercise 5.4 Suppose $f(x)$ is given by a power series $\sum_{n=0}^{\infty} a_n x^n$, with radius of convergence $R > 0$. For $m \in \mathbb{N}$ show that

$$f(x) = \sum_{n=0}^{m-1} a_n x^n + O(x^m)$$

as $x \to 0$.

The second order notation, referred to as "little o," is to write

$$f(x) = o(\phi(x)) \quad \text{as } x \to \alpha \tag{5.5}$$

to signify that

$$\lim_{x \to \alpha} \frac{f(x)}{\phi(x)} = 0.$$

For example, we can paraphrase the function limit $f(x) \to y$ as $x \to \alpha$ by writing that

$$f(x) = y + o(1).$$

An error of $o(1)$ simply indicates a term that goes to zero in the limit.

The final notation to introduce here describes a more precise relationship between functions in the limit. We say that f is *asymptotic* to ϕ, written as

$$f(x) \sim \phi(x) \quad \text{as } x \to \alpha, \tag{5.6}$$

if

$$\lim_{x \to \alpha} \frac{f(x)}{\phi(x)} = 1.$$

For example, from (5.4) we can write

$$\sin x \sim x \quad \text{as } x \to 0,$$

if we are only interested in the limiting behavior. Unlike the order notations previously introduced, the definition (5.6) is clearly an equivalence relation. Since $\sim$ can be used

5.1 Limits and Continuity

for more general equivalence relations, the use of $\sim$ to denote an asymptotic needs to be indicated explicitly, if it is not clear from context.

Exercise 5.5 Show that $f(x) \sim \phi(x)$ as $x \to \alpha$ if and only if

$$f(x) = \phi(x) + o(\phi(x)).$$

5.1.2 Continuous Functions

According to the general definition from Sect. 4.1, a function $f : I \to \mathbb{R}$ is continuous at a point $x_0 \in I$ if for any $\varepsilon > 0$ there exists $\delta > 0$ such that

$$|f(x) - f(x_0)| < \varepsilon$$

for all $x \in I$ with $|x - y| < \delta$. In terms of the order notation introduced in Sect. 5.1.1, this could be written more succinctly as

$$f(x) = f(x_0) + o(1) \text{ as } x \to x_0.$$

There are some other equivalent ways to describe continuity. By Exercise 4.6 we could say that f is continuous at x_0 if and only if $f(y_n) \to f(x_0)$ for every convergent sequence $y_n \to x_0$ in I. For continuity on the full interval, the topological formulation given in Theorem 4.8 also applies: f is continuous on I if and only if $f^{-1}(U)$ is open relative to I for every open set $U \subset \mathbb{R}$.

By the same basic arguments used for Lemma 5.1, we can see that continuity is preserved under algebraic operations:

Lemma 5.6 *If f and g are continuous real-valued functions, then so are $f + g$ and fg.*

We have already noted that monomials are continuous in Example 4.1, so Lemma 5.6 implies that all polynomials are continuous. That simple argument does not extend to power series, because linearity allows only finite linear combinations, but we have already dealt with continuity for power series in Example 4.16. We have thus already established continuity for the most common classes of functions: polynomials, exponentials, trigonometric, etc. These cases are presumably already familiar from calculus.

Other properties of continuous functions have already been covered in the general case in Sect. 4.1. In Exercise 4.14 we saw that continuous functions map connected sets to connected sets. And in Theorem 3.20 we showed that the connected subsets of $\mathbb{R}$ are precisely the intervals. Together, these results yield the following:

Theorem 5.7 (**intermediate value**) *If f is a real-valued continuous function defined on an interval I, then $f(I)$ is an interval.*

Exercise 5.8 For a function $f : [a, b] \to \mathbb{R}$, define the *graph*

$$\Gamma(f) := \{(x, f(x)) : x \in [a, b]\} \subset \mathbb{R}^2.$$

Prove that f is continuous on $[a, b]$ if and only if $\Gamma(f)$ is compact.

At a point of discontinuity we sometimes want to compare separate limits from the two sides. Suppose that f is defined on an open interval (a, b). For $x_0 \in (a, b)$. The left and right limits,

$$\lim_{x \to x_0^+} f(x), \quad \lim_{x \to x_0^-} f(x),$$

are defined by restricting to the subdomains with (x_0, b) and (a, x_0), respectively. The left and right limits may exist in cases where the full limit does not. For example, if $\lfloor x \rfloor$ denotes the greatest integer function, then for $n \in \mathbb{Z}$,

$$\lim_{x \to n^+} \lfloor x \rfloor = n, \quad \lim_{x \to n^-} \lfloor x \rfloor = n - 1.$$

A singularity of this type, where the left and right limits exist but are not equal, is called a *jump discontinuity*.

A real function is *increasing* if $f(x) \le f(y)$ for all $x < y$ and *decreasing* if $f(x) \ge f(y)$. The function is called *monotonic* if it is either increasing or decreasing. For each jump discontinuity, we can choose a rational number that lies strictly between the left and right limits. If the function is monotonic, then the rational numbers so assigned will be distinct. Since there are only countably many rationals to choose from, this implies the following:

Exercise 5.9 For a monotonic function $f : I \to \mathbb{R}$, prove that the set of points of discontinuity is at most countable and all discontinuities are jumps.

5.2 Differentiation

A function $f : I \to \mathbb{R}$ is *differentiable* at $x \in I$ if the limit

$$f'(x) := \lim_{t \to 0} \frac{f(x + t) - f(x)}{t}, \tag{5.7}$$

called the *derivative*, exists. We say that f is differentiable on I if $f'(x)$ exists for all $x \in I$. The derivative function is also commonly written as

$$f' = \frac{df}{dx},$$

5.2 Differentiation

Fig. 5.1 A function and its tangent line approximation at x_0

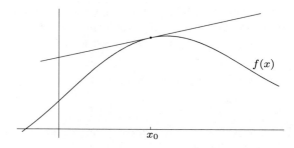

to reflect its definition as an infinitesimal limit of difference quotients.

In terms of our order notation, the definition of $f'(x_0)$ is equivalent to the statement

$$f(x) = f(x_0) + f'(x_0)(x - x_0) + o(x - x_0), \tag{5.8}$$

as $x \to x_0$. Existence of the derivative is thus equivalent to an error estimate on the linear approximation given by the tangent line, as illustrated in Fig. 5.1.

We assume that basic derivative calculations are already familiar to the reader, but we will go through a few examples just to illustrate the definition.

Example 5.10 To differentiate the monomial function $f(x) = x^n$ for $n \in \mathbb{N}$, we rewrite the identity (4.2) as

$$\frac{y^n - x^n}{y - x} = y^{n-1} + y^{n-1}x + \cdots + x^{n-1}.$$

Taking $y \to x$ thus gives $f'(x) = nx^{n-1}$.   ◇

Example 5.11 Consider the exponential function as defined in Example 2.15, restricted to a real variable. By the addition formula of Lemma 2.19, the difference quotient can be factored:

$$\frac{e^{x+t} - e^x}{t} = \frac{e^t - 1}{t} e^x. \tag{5.9}$$

From the power series expansion we can see that

$$\lim_{t \to 0} \frac{e^t - 1}{t} = 1.$$

Hence the limit $t \to 0$ in (5.9) exists and yields the familiar formula:

$$\frac{d}{dx} e^x = e^x. \tag{5.10}$$

◇

The basic properties of the derivative are presumably well-known to the reader, but it is worth recalling the proofs. These are essentially direct applications of the algebraic identities for function limits proven in Lemma 5.1.

Exercise 5.12 Show that a real function that is differentiable at a point is also continuous at that point.

Exercise 5.13 Let f and g be differentiable functions $I \to \mathbb{R}$. Prove that $f + g$ and fg are differentiable, with
$$(f + g)' = f' + g', \qquad (fg)' = f'g + fg'.$$

Exercise 5.13 implies in particular that the set of differentiable functions is a vector space, and the derivative operator is a linear map.

The last basic fact we need to review is the chain rule, which has a slightly trickier proof than the exercises above.

Theorem 5.14 *Suppose that* $f : I \to \mathbb{R}$ *and* $g : J \to \mathbb{R}$ *are differentiable functions defined on intervals such that* $f(I) \subset J$. *Then the composition* $g \circ f$ *is differentiable and*
$$(g \circ f)' = (g' \circ f) f'.$$

Proof Fix a point $x \in I$, and let $y = f(x)$. For s small enough that $y + s \in J$, define the function
$$\eta(s) := \begin{cases} \frac{g(y+s)-g(y)}{s}, & s \neq 0, \\ g'(y), & s = 0, \end{cases}$$
which is continuous because g is differentiable. The difference quotient for $g \circ f$ can then be written
$$\frac{g(f(x+t)) - g(f(x))}{t} = \eta(f(x+t) - f(x)) \left(\frac{f(x+t) - f(x)}{t} \right) \quad \text{for } t \neq 0. \quad (5.11)$$

Since η and f are continuous,
$$\lim_{t \to 0} \eta(f(x+t) - f(x)) = \eta(0) = g'(y).$$

We can thus take the limit $t \to 0$ in (5.11) to obtain
$$(g \circ f)'(x) = g'(y) f'(x).$$
□

A function $f : I \to \mathbb{R}$ is *convex* if for all points $x, y \in I$ and $t \in [0, 1]$,

5.2 Differentiation

$$(1-t)f(x) + tf(y) \geq f((1-t)x + ty). \tag{5.12}$$

The left-hand side of (5.12) is the interpolation between the values of f at x and y, while the right is f evaluated at the corresponding interpolated point. Thus convexity means that in the graph the secant lines lie above the curve. In calculus such functions are also called *concave up*. For a differentiable function, we can express this in terms to the tangent line approximation given by (5.8).

Exercise 5.15 Suppose f is a differentiable function on an open interval I. Prove that f is convex if and only if for each $x_0 \in I$ the graph of f lies above the tangent line at x_0, meaning

$$f(x) \geq f(x_0) + f'(x_0)(x - x_0) \tag{5.13}$$

for all $x \in I$.

Example 5.16 The exponential function is convex, as illustrated in Example 5.2. While the inequality (5.12) is not so obvious in this case, the tangent condition (5.13) is easily checked. By the addition rule (Lemma 2.19) and the derivative formula (5.10), it suffices to show that

$$e^x \geq 1 + x \tag{5.14}$$

for all $x \in \mathbb{R}$. For $x \geq 0$ this follows from the positivity of the exponential power series coefficients, and the inequality is trivial for $x \leq -1$ because $e^x > 0$. For $-1 < x < 0$, we can compare the exponential and geometric series,

$$e^{-x} = \sum_{n=0}^{\infty} \frac{(-x)^n}{n!} \leq \sum_{n=0}^{\infty} (-x)^n = (1+x)^{-1}.$$

Taking the reciprocal gives (5.14) for $x \in (-1, 0)$. ◇

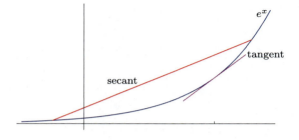

Fig. 5.2 The exponential as an example of a convex function

5.2.1 Differentiation of Power Series

In Sect. 2.4 we defined exponential and trigonometric functions in terms of power series. Since most of the important real functions admit a local power series expansion, it is useful to study the differentiability of power series in general. We are considering only real derivatives here, but similar arguments apply to complex derivatives.

Consider a real function f defined as

$$f(x) = \sum_{k=0}^{\infty} c_k x^k \tag{5.15}$$

for $c_k \in \mathbb{R}$. We know from Theorem 2.18 that $f(x)$ converges on an interval $(-R, R)$ with R given by the formula (2.9). We will always assume that $R > 0$.

The differentiation of a single term from the series is covered by Example 5.10, and by linearity this familiar formula applies to polynomials. It turns out that we can differentiate the power series (5.15) by applying the same rule to each term. However, since the series is defined as a limit, this does not immediately follow from the polynomial formula.

Theorem 5.17 *The function f defined by the power series (5.15) is differentiable on $(-R, R)$, with*

$$f'(x) = \sum_{k=1}^{\infty} k c_k x^{k-1}.$$

Proof Define f_n as the partial sum

$$f_n(x) := \sum_{k=0}^{n} c_k x^k.$$

This is just a polynomial and so we already know that $f'_n = g_n$, where

$$g_n(x) := \sum_{k=1}^{n} k c_k x^{k-1}.$$

The calculation (4.6) from Example 4.16 shows that the g_n are partial sums of a power series

$$g(x) := \sum_{k=1}^{\infty} k c_k x^{k-1}$$

that converges on the same interval $(-R, R)$.

For $x \in (-R, R)$ our goal is to show that $f'(x) = g(x)$. Replacing f by f_n in the difference quotient gives

5.2 Differentiation

$$\left|\frac{f(y)-f(x)}{y-x}-g(x)\right| \le \left|\frac{f_n(y)-f_n(x)}{y-x}-g_n(x)\right| + |g_n(x)-g(x)| \\ + \left|\frac{1}{y-x}\sum_{k=n+1}^{\infty} c_k(y^k-x^k)\right| \qquad (5.16)$$

for $y \ne x$. Let $\varepsilon > 0$. For the third term on the right we estimate as in (4.5) from Example 4.16, using

$$\frac{1}{y-x}\sum_{k=n+1}^{\infty} c_k(y^k-x^k) = \sum_{k=n+1}^{\infty} c_k(y^{k-1}+\cdots+x^{k-1})$$

For r such that $|x| < r < R$ we can assume that $|y| < r$ also. Then

$$\left|\frac{1}{y-x}\sum_{k=n+1}^{\infty} c_k(y^k-x^k)\right| \le \sum_{k=n+1}^{\infty} k|c_k|r^{k-1}.$$

Since the sum on the right is convergent for $r < R$, for n sufficiently large we have

$$\left|\frac{1}{y-x}\sum_{k=n+1}^{\infty} c_k(y^k-x^k)\right| < \varepsilon$$

for $|y| < r$. Furthermore, since $g_n \to g$ on this interval, we can assume that n is chosen so that

$$|g_n(x)-g(x)| < \varepsilon.$$

For this value of n, (5.16) reduces to

$$\left|\frac{f(y)-f(x)}{y-x}-g(x)\right| \le \left|\frac{f_n(y)-f_n(x)}{y-x}-g_n(x)\right| + 2\varepsilon. \qquad (5.17)$$

By Lemma 5.2, we can take the lim sup as $y \to x$ on both sides of (5.16), and since $f'_n = g_n$ the first term on the right drops out in the limit. This leaves

$$\limsup_{y \to x}\left|\frac{f(y)-f(x)}{y-x}-g(x)\right| \le 2\varepsilon.$$

Since ε was arbitrary, this proves that the lim sup is equal to zero, and hence by Lemma 5.2,

$$\lim_{y \to x}\left|\frac{f(y)-f(x)}{y-x}-g(x)\right| = 0.$$

We have thus shown that $f'(x) = g(x)$. $\square$

Note that in the final stage of the proof we could not simply take the limit as $y \to x$ on both sides of (5.17), because the limit on the left is not yet known to exist. The lim sup provides a good way to handle such situations.

5.2.2 Higher Derivatives

If f is a differentiable function and f' is also differentiable at x, then the second derivative is defined as

$$f''(x) := \frac{d}{dx} f'(x).$$

Exercise 5.18 Suppose that f is twice differentiable on an open interval I and convex. Prove that $f'' > 0$ at all points.

Higher derivatives, if they exist, are denoted as

$$f^{(n)}(x) := \frac{d^n}{dx^n} f(x).$$

For a function represented by power series, the differentiation formula of Theorem 5.17 can be applied repeatedly, since the radius of convergence stays fixed. Hence that theorem has the following:

Corollary 5.19 *Suppose that $f(x)$ is defined near x_0 by the power series*

$$f(x) = \sum_{k=0}^{\infty} c_k (x - x_0)^k,$$

with radius $R > 0$. Then f differentiable to all orders and the coefficients satisfy

$$c_k = \frac{1}{k!} f^{(k)}(x_0). \qquad (5.18)$$

The converse to Corollary 5.19 is false. Some infinitely differentiable functions cannot be represented as a power series.

Example 5.20 For $x \in \mathbb{R}$ let

$$f(x) := \begin{cases} e^{-1/x^2}, & x \neq 0, \\ 0, & x = 0, \end{cases}$$

5.3 The Mean Value Theorem

Fig. 5.3 The graph of the function from Example 5.20

as pictured in Fig. 5.3. Away from $x = 0$ this can be differentiated by the standard rules. And we can deduce that $f^{(n)}(0) = 0$ for all n from the fact that

$$\lim_{x \to 0} \frac{e^{-1/x^2}}{x^n} = 0$$

for all n. If f admitted a power series expansion at $x = 0$, then the coefficients would be given by (5.18). Since these are all zero, there is no such expansion. ◇

We say that f is C^m, or "m-times continuously differentiable", if the derivatives $f^{(n)}$ exist for all $n \leq m$ and $f^{(m)}$ is also continuous. A function is said to be C^∞, or *infinitely differentiable* if derivatives exist to all orders. Corollary 5.19 shows that power series define C^∞ functions.

5.3 The Mean Value Theorem

Many theoretical results involving the derivative are established using the mean value theorem. To set up the proof we first consider extremal points. A real-valued function f has a *local maximum* at x_0 if there exists $\delta > 0$ so that

$$f(x) \leq f(x_0) \quad \text{for } |x - x_0| < \delta. \tag{5.19}$$

Similarly, f has a *local minimum* at x_0 if there exists $\delta > 0$ such that

$$f(x) \geq f(x_0) \quad \text{for } |x - x_0| < \delta. \tag{5.20}$$

A point that is either a local minimum or maximum is called a *local extremum*.

Exercise 5.21 Suppose f is a differentiable real-valued function defined on an open interval I. If $x_0 \subset I$ is a local extremum, show that $f'(x_0) = 0$.

Theorem 5.22 (**mean value**) *Suppose $f : [a, b] \to \mathbb{R}$ is a continuous function and that f is differentiable on (a, b). There exists $t \in (a, b)$ such that*

Fig. 5.4 The original function f is modified to create h, flattening the secant line between a and b

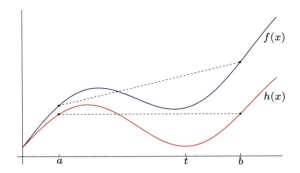

$$\frac{f(b) - f(a)}{b - a} = f'(t).$$

Proof For convenience let m denote the slope of the secant line,

$$m := \frac{f(b) - f(a)}{b - a}.$$

Then the function $h(x) = f(x) - mx$ satisfies $h(a) = h(b)$, as illustrated in Fig. 5.4. By the extreme value theorem (Theorem 4.12), h attains a maximum and minimum value in $[a, b]$. If these both occur at endpoints, then h is constant and $f' = m$ at all points. If h is not constant then at least one extremum must occur at an interior point. If we let t denote such a point, then Exercise 5.21 implies $h'(t) = 0$, which yields $f'(t) = m$. □

One immediate application is a relation between the monotonicity of a function and the sign of its derivative. It is already clear that $f' \geq 0$ for a differentiable increasing function, because the difference quotients in (5.7) are positive. Similarly, a decreasing function has $f' \geq 0$. The mean value theorem clearly implies the converse statement:

Corollary 5.23 *A differentiable function $f : I \in \mathbb{R}$ is increasing if and only if $f' \geq 0$ on I and decreasing if and only if $f' \leq 0$.*

We can carry the conclusion of Corollary 5.23 further if the derivative function has no zeros.

Theorem 5.24 (inverse function) *On an open interval I suppose that $f : I \to \mathbb{R}$ is differentiable with $f'(t) \neq 0$ for all $t \in I$. Then f is injective and the inverse function $f^{-1} : f(I) \to I$ is differentiable. At the point $y = f(x)$,*

$$(f^{-1})'(y) = \frac{1}{f'(x)}.$$

5.3 The Mean Value Theorem

Proof We can assume that $f' > 0$ by replacing f by $-f$ if necessary. Theorem 5.22 implies that f is strictly increasing and therefore injective. This defines f^{-1} as a function on $J = f(I)$. We know that f maps intervals to intervals by Theorem 5.7, and from the fact that f is strictly increasing we can deduce that f maps open intervals to open intervals. It follows that J is an open interval and f^{-1} is continuous.

To see that f^{-1} is differentiable, let $y = f(x)$ and $y_0 = f(x_0)$ for $x, x_0 \in I$. By continuity, $y \to y_0$ as $x \to x_0$. Since $f'(x_0) \neq 0$ we can replace the variables to see that

$$\frac{1}{f'(x_0)} = \lim_{x \to x_0} \frac{x - x_0}{y - y_0} = \lim_{y \to y_0} \frac{x - x_0}{y - y_0} = (f^{-1})'(y_0).$$

$\square$

Example 5.25 The logarithm is defined as the inverse of the exponential function. For $x \geq 0$ it is evident that $e^x > 0$ from the power series expansion, and it then follows from the addition formula (2.11) that e^x is strictly positive for all x. By (5.10) the derivative is strictly positive, and so Theorem 5.24 gives us an inverse function $\log : (0, \infty) \to \mathbb{R}$ such that

$$y = e^x \iff x = \log y.$$

The inverse formula for the derivative reduces in this case to

$$\frac{d}{dy} \log y = \frac{1}{e^x} = \frac{1}{y}. \tag{5.21}$$

$\diamond$

Now that the logarithm has been introduced, we can provide the proof of Young's inequality, which was claimed in (3.5). For $p, q > 1$ such that $1/p + 1/q = 1$,

$$xy \leq \frac{x^p}{p} + \frac{x^q}{q} \tag{5.22}$$

for all $x, y > 0$. Recall that the exponential function was shown to be convex in Example 5.16. Using the relation between p and q to interpolate, the convexity inequality (5.12) gives

$$\frac{1}{p} e^a + \frac{1}{q} e^b \leq \exp\left(\frac{a}{p} + \frac{b}{q}\right).$$

for all $a, b \in \mathbb{R}$. If we set $a = p \log x$ and $b = q \log y$ then this reduces to (5.22).

5.3.1 Taylor Approximation

Another important application of the mean value theorem is Taylor's theorem on the approximation of a differentiable function by polynomials. The coefficients of these polynomials agree with the power series formula (5.18), but the Taylor approximation does not require a full power series expansion.

Theorem 5.26 (Taylor approximation) *Suppose f is a C^m function defined in a neighborhood of x_0, and define the polynomial*

$$p_m(x) := \sum_{k=0}^{m} \frac{1}{k!} f^{(k)}(x_0)(x - x_0)^k.$$

Then, as $x \to x_0$,

$$f(x) = p_m(x) + o(|x - x_0|^m).$$

Proof Set

$$g(x) = f(x) - p_m(x),$$

so that $g^{(k)}(x_0) = 0$ for $k = 0, \ldots, m$. The mean value theorem applied to $g(x) - g(x_0)$ gives x_1 between x_0 and x so that

$$g(x) = g'(x_1)(x - x_0).$$

Then the mean value theorem for $g'(x_1) - g'(x_0)$ gives x_2 between x_0 and x_1 so that

$$g'(x_1) = g''(x_2)(x_1 - x_0),$$

and so on. After continuing this process, we obtain $x_1, \ldots, x_m$ such that

$$g^{(n-1)}(x_{n-1}) = g^{(n)}(x_n)(x_{n-1} - x_0).$$

Combining all these expressions gives

$$g(x) = g^{(m)}(x_m)(x - x_0)(x_1 - x_0) \ldots (x_{n-1} - x_0).$$

Since the points all lie between x and x_0, we have the bound

$$|g(x)| \leq |g^{(m)}(x_m)||x - x_0|^m.$$

Finally, the assumption that $g^{(m)}$ is continuous implies that $g^{(m)}(x_m) \to 0$ as $x \to 0$, since x_m lies between x_0 and x. This proves that

$$\lim_{x \to x_0} \frac{|g(x)|}{|x - x_0|^m} = 0$$

5.3 The Mean Value Theorem

which is the claimed result. □

We can improve the error estimate in Taylor's approximation if we have one extra derivative. For the $m = 0$ case this improvement is seen in the original mean value theorem. If f is differentiable in neighborhood of x_0, then by Theorem 5.22

$$f(x) = f(x_0) + f'(t)(x - x_0),$$

for some t between x_0 and x. If f' is bounded near x_0, then this gives an error term $O(x - x_0)$, whereas Theorem 5.26 gives only $o(1)$. This idea can be extended to higher derivatives. Here for example is the quadratic statement.

Theorem 5.27 *Suppose that f is twice differentiable on an open interval I. Given $x, x_0 \in I$,*

$$f(x) = f(x_0) + f'(x_0)(x - x_0) + \frac{1}{2} f''(t)(x - x_0)^2 \tag{5.23}$$

for some t between x_0 and x.

Proof Let $p_1(t) = f(x_0) + f'(x_0)(t - x_0)$ and set

$$c = \frac{f(x) - p_1(x)}{(x - x_0)^2}.$$

Then the function

$$g(t) := f(t) - p_1(t) - c(t - x_0)^2$$

satisfies $g(x) = g(x_0) = 0$ as well as $g'(x_0) = 0$. The mean value theorem gives a point x_1 between x_0 and x such that $g'(x_1) = 0$. Applying it again gives a point t between x_0 and x_1 such that $g''(t) = 0$. Since

$$g''(t) = f''(t) - 2c,$$

this proves (5.23). □

The quadratic error term in Theorem 5.27 yields an immediate converse to the result of Exercise 5.18. That is, if $f'' > 0$ at all points of an interval I, then (5.23) gives

$$f(x) \geq f(x_0) + f'(x_0)(x - x_0)$$

for all $x, x_0 \in I$. This implies that f is convex by Exercise 5.15.

5.3.2 L'Hôpital's Rule

Limits of ratios of functions may have an indeterminate form, where the ratio appears to approach 0/0 or ∞/∞. L'Hôpital's rule gives an effective tool for evaluating such limits. The result can be applied to a variety of situations. The proof uses a more general version of the mean value theorem, which we state first as a lemma.

Lemma 5.28 *Suppose f and g are continuous functions $[a, b] \to \mathbb{R}$ which are differentiable on (a, b). There exists $t \in (a, b)$ such that*

$$[f(b) - f(a)]g'(t) = [g(b) - g(a)]f'(t).$$

Proof We set
$$h(x) := [f(b) - f(a)]g(t) - [g(b) - g(a)]f(t),$$
so that $h(a) = h(b)$. Theorem 5.22 then implies that $h'(t) = 0$ for some $t \in (a, b)$. □

Theorem 5.29 *Suppose f and g are real differentiable functions on the interval (a, ∞), with $g'(x) \neq 0$ for all $x > a$. If $f(x) \to 0$ and $g(x) \to 0$ as $x \to \infty$, and the limit*

$$\lim_{x \to \infty} \frac{f'(x)}{g'(x)} = \alpha \tag{5.24}$$

exists with $\alpha \in \mathbb{R}_\infty$, then

$$\lim_{x \to \infty} \frac{f(x)}{g(x)} = \alpha.$$

Proof Since $g' \neq 0$, the function g is strictly monotonic by Theorem 5.24, and so $g(x) \to 0$ as $x \to \infty$ implies $g(x) \neq 0$ for all $x > a$.

Suppose $\alpha < \infty$ and c be a real number with $c > \alpha$. By the hypothesis (5.24) there exists $m > a$ so that
$$\frac{f'(t)}{g'(t)} < c \quad \text{for } t > m. \tag{5.25}$$

For $y > x > a$, since g is strictly monotonic, the result of Lemma 5.28 can be rearranged to give
$$\frac{f(x) - f(y)}{g(x) - g(y)} = \frac{f'(t)}{g'(t)}$$

for some $t \in (x, y)$. Thus, by (5.25),
$$\frac{f(x) - f(y)}{g(x) - g(y)} < c \quad \text{for all } y > x > m.$$

Taking $y \to \infty$ gives

$$\frac{f(x)}{g(x)} < c \quad \text{for all } x > m,$$

since $f(y) \to 0$, $g(y) \to 0$, and $g(x) \neq 0$. Since $c > \alpha$ was arbitrary, we can conclude from this that

$$\limsup_{x \to \infty} \frac{f(x)}{g(x)} \leq \alpha. \tag{5.26}$$

We assumed $\alpha < \infty$ for the derivation, but (5.26) obviously still holds by default if $\alpha = \infty$.

With the same strategy, by taking $c < \alpha$ we can argue that

$$\liminf_{x \to \infty} \frac{f(x)}{g(x)} \geq \alpha,$$

and this completes the proof. □

Other forms of l'Hôpital's rule can be deduced from Theorem 5.29. For example, we can switch the limit to $x \to 0$.

Corollary 5.30 *Suppose f and g are real differentiable functions on the interval $(0, b)$, with $g'(x) \neq 0$ for all $x \in (0, b)$. If $f(x) \to 0$ and $g(x) \to 0$ as $x \to 0$, and the limit*

$$\lim_{x \to 0} \frac{f'(x)}{g'(x)} = \alpha$$

exists with $\alpha \in \mathbb{R}_\infty$, then

$$\lim_{x \to 0} \frac{f(x)}{g(x)} = \alpha.$$

Proof Define $h_1(y) = f(1/y)$ and $h_2(y) = g(1/y)$. Then by the chain rule,

$$\frac{h'_1(y)}{h'_2(y)} = \frac{f'(1/y)}{g'(1/y)}.$$

The limit of this ratio as $y \to \infty$ is α, and so Theorem 5.29 gives

$$\lim_{y \to \infty} \frac{f(1/y)}{g(1/y)} = \alpha.$$

□

The ∞/∞ case can be handled the same way, by applying Theorem 5.29 to the functions $1/f$ and $1/g$.

5.4 Integration

In this section we will consider the standard Riemann definition of the integral for a bounded function on a closed interval. We will generally restrict our attention to continuous functions, which is sufficient for the applications we can develop here. The full generalization of the concept of integrability requires the Lebesgue measure theory, which lies beyond the scope of this text.

A *partition* of a bounded interval $[a, b]$ is a finite set of points $P = \{x_0, \ldots, x_n\}$ with

$$a = x_0 \leq x_1 \leq \cdots \leq x_n = b.$$

For $f : [a, b] \to \mathbb{R}$ bounded, we can associate to P the *upper sum*

$$S^+(f, P) := \sum_{j=1}^{n} (x_j - x_{j-1}) \sup_{[x_{j-1}, x_j]} f$$

and *lower sum*

$$S^-(f, P) := \sum_{j=1}^{n} (x_j - x_{j-1}) \inf_{[x_{j-1}, x_j]} f,$$

as illustrated in Fig. 5.5. It is clear from the definitions that

$$S^-(f, P) \leq S^+(f, P).$$

A bounded function f is *Riemann integrable* if

$$\inf_P S^+(f, P) = \sup_P S^-(f, P),$$

and $\int_a^b f$ is defined as this common value. When there is a need to indicate the integration variable, we write this as

$$\int_a^b f = \int_a^b f(x)\,dx.$$

A partition P_2 is called a *refinement* of P_1 if $P_1 \subset P_2$. In other words, P_2 contains all of the break points of P_1, plus a finite number of new subdivisions.

Lemma 5.31 *If P_2 is a refinement of P_1, then*

$$S^+(f, P_2) \leq S^+(f, P_1), \quad S^-(f, P_2) \geq S^-(f, P_1).$$

Proof Since P_2 differs from P_1 by finitely many points, it suffices to consider the addition of a single point. Suppose $P_1 = \{x_1, \ldots, x_n\}$ and $P_2 = P_1 \cup \{y\}$ where $c \in (x_j, x_{j+1})$. The supremum of f over either subinterval of $[x_j, x_{j+1}]$ is less than the supremum over the full interval. This implies

5.4 Integration

Fig. 5.5 The upper and lower sums correspond to approximation by step functions

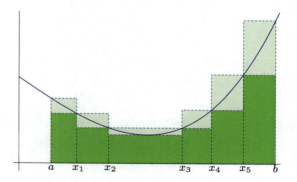

$$(c - x_j) \sup_{[x_j, c]} f + (x_{j+1} - c) \sup_{[c, x_{j+1}]} f \leq (x_{j+1} - x_j) \sup_{[x_j, x_{j+1}]} f,$$

which shows that $S^+(f, P_2) \leq S^+(f, P_1)$. A similar argument applies to S^-. □

Two partitions P_1 and P_2 have a common refinement given by

$$P = P_1 \cup P_2.$$

Lemma 5.31 allows us to deduce that

$$S^+(f, P_1) \geq S^-(f, P_2)$$

for any partitions P_1, P_2. Taking the infimum over P_1 and the supremum over P_2 then shows that

$$\inf_P S^+(f, P) \geq \sup_P S^-(f, P). \tag{5.27}$$

Theorem 5.32 *A continuous function $f : [a, b] \to \mathbb{R}$ is Riemann integrable.*

Proof Let P_n denote the regular partition with

$$x_j = a + \frac{b-a}{n} j.$$

Since $[a, b]$ is compact, f is uniformly continuous by Exercise 4.17. Hence, given $\varepsilon > 0$ there exists $\delta > 0$ so that

$$|f(x) - f(y)| < \varepsilon, \quad \text{for } |x - y| < \delta.$$

Thus, for $n > (b-a)/\delta$, the sup and inf of f on each interval of P_n differ by at most ε. This implies that

$$S^+(f, P_n) - S^-(f, P_n) \leq (b-a)\varepsilon,$$

which gives
$$\inf_P S^+(f, P) - \sup_P S^-(f, P) \le (b-a)\varepsilon.$$

Since the left side is positive, by (5.27), and ε was arbitrary, we conclude that
$$\inf_P S^+(f, P) = \sup_P S^-(f, P)$$

$\square$

It is easy to adapt the proof of Theorem 5.32 to allow f to have finitely many points of discontinuity, and in fact the result can be improved even further. However, the strongest generalization requires the Lebesgue definition of the integral via measure theory, which we will not get into here.

Some basic properties of the Riemann integral follow immediately from the definition. The most fundamental of these is linearity. For any partition P and $c \ge 0$ it is clear that
$$S^\pm(cf, P) = cS^\pm(f, P),$$
while for $c < 0$ the right side would be $cS^\mp(f, P)$. It follows that
$$\int_a^b cf = c \int_a^b f$$
for $c \in \mathbb{R}$. In conjunction with the following result, this shows that integration is a linear operation.

Lemma 5.33 *If f_1 and f_2 are integrable functions on $[a, b]$, then so is $f_1 + f_2$, with*
$$\int_a^b (f_1 + f_2) = \int_a^b f_1 + \int_a^b f_2.$$

Proof For any partition P the triangle inequality implies that
$$S^-(f_1, P) + S^-(f_2, P) \le S^+(f_1 + f_2, P)$$
and
$$S^+(f_1 + f_2, P) \le S^+(f_1, P) + S^+(f_2, P).$$
Given $\varepsilon > 0$ we can find a partition P such that
$$S^+(f_j, P) - S^-(f_j, P) \le \varepsilon$$
for $j = 1, 2$, by taking a common refinement of the partitions for each function. For this P we have

5.4 Integration

$$S^-(f_1 + f_2, P) \geq \int_a^b f_1 + \int_a^b f_2 - 2\varepsilon$$

and

$$S^+(f_1 + f_2, P) \leq \int_a^b f_1 + \int_a^b f_2 + 2\varepsilon.$$

Since ε was arbitrary, we conclude that $f_1 + f_2$ is integrable and

$$\int_a^b (f_1 + f_2) = \int_a^b f_1 + \int_a^b f_2.$$

$\square$

Another basic but important property is the concatenation of integrals over adjacent intervals. We will omit the very straightforward proof.

Lemma 5.34 *Suppose that f is integrable on $[a, b]$. For $a < c < b$,*

$$\int_a^b f = \int_a^c f + \int_c^b f.$$

Because of this additive property, it makes sense to define, for $a < b$,

$$\int_b^a f := -\int_a^b f.$$

The final property we wish to highlight is *positivity*. If f is a positive function then clearly the sums $S^\pm(f, P)$ are positive for any partition P. Therefore, assuming integrability,

$$f \geq 0 \text{ on } [a, b] \implies \int_a^b f \geq 0. \tag{5.28}$$

The combination of linearity and positivity implies the monotonicity of the integral:

$$f \leq g \text{ on } [a, b] \implies \int_a^b f \leq \int_a^b g. \tag{5.29}$$

Since $\pm f \leq |f|$, monotonicity yields the basic integral estimate:

$$\left| \int_a^b f \right| \leq \int_a^b |f|. \tag{5.30}$$

Since f is bounded, we can extract its estimate from the integral to obtain

$$\left| \int_a^b f \right| \leq (b - a) \sup_{x \in [a,b]} |f(x)|. \tag{5.31}$$

Exercise 5.35 Suppose that f_n is integrable on $[a, b]$ for each n and $f_n \to f$ uniformly. Prove that f is integrable and
$$\lim_{n \to \infty} \int_a^b f_n = \int_a^b f.$$

Exercise 5.36 Let V be the vector space of continuous functions $[0, 1] \to \mathbb{R}$ and define
$$\|f\| := \int_a^b |f|$$
for $f \in V$. Show that $\|\cdot\|$ defines a norm on V.

5.4.1 Fundamental Theorem of Calculus

The fundamental theorem says that differentiation and integration are inverse operations, in a certain sense. There are two parts to the statement, depending on the order of the operations. First, we consider differentiation of the integral.

Theorem 5.37 *Suppose that f is Riemann integrable on the interval $[a, b]$. The function*
$$F(x) := \int_a^x f$$
is Lipschitz continuous on $[a, b]$. If f is continuous at a point x then F is differentiable at x with
$$F'(x) = f(x).$$

Proof For $x, y \in [a, b]$, Lemma 5.34 implies that
$$F(y) - F(x) = \int_x^y f. \tag{5.32}$$

Lipschitz continuity then follows from the estimate (5.30) which gives
$$|F(y) - F(x)| \leq M|y - x|,$$
where $M = \sup_{[a,b]} |f|$. (M is finite under the assumption that f is Riemann integrable.)

Now assume that f is continuous at x. By (5.32) we can write
$$\frac{F(x+t) - F(x)}{t} - f(x) = \frac{1}{t} \int_x^{x+t} [f(y) - f(x)] \, dy.$$

Applying (5.30) gives the estimate

$$\left|\frac{F(x+t)-F(x)}{t} - f(x)\right| \le \sup_{y\in[x,x+t]} |f(y)|$$

assuming $x+t \in [a,b]$. Continuity implies that the right-hand side is $o(1)$ as $t \to 0$, which proves that $F'(x)$ exists and equals $f(x)$. □

The second part of the fundamental theorem involves integration of the derivative.

Theorem 5.38 *Suppose f is differentiable on $[a,b]$ and f' is integrable. Then*

$$\int_a^b f' = f(b) - f(a).$$

Proof Given a partition $P = \{x_0, \ldots, x_n\}$, applying the mean value theorem to each subinterval gives $t_j \in (x_{j-1}, x_j)$ such that

$$f(x_j) - f(x_{j-1}) = f'(t_j)(x_j - x_{j-1}).$$

Summing over j then gives

$$f(b) - f(a) = \sum_{j=1}^n f'(t_j)(x_j - x_{j-1}).$$

This shows that

$$S^-(f', P) \le f(b) - f(a) \le S^+(f', P).$$

If f is integrable then taking the supremum over P in the first inequality and the infimum in the second proves the result. □

Exercise 5.39 Let f_n be a C^1 function on the interval $[a,b]$ for each $n \in \mathbb{N}$. Suppose that there are functions f, g such that $f_n \to f$ pointwise and $f'_n \to g$ uniformly. Prove that f is C^1 and $f' = g$.

5.5 Picard Iteration

In this final section we will illustrate how metric space tools can be applied to establish the existence of solutions of an ordinary differential equation. The proof involves a iteration technique introduced by Émile Picard in 1893.

Fig. 5.6 A continuous slope field $F(t, u)$ with a possible solution of the corresponding differential equation

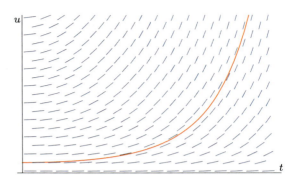

On an interval $I \subset \mathbb{R}$ we consider an equation of the form

$$u'(t) = F(t, u(t)), \qquad u(t_0) = u_0 \tag{5.33}$$

where F is a continuous function $I \times \mathbb{R} \to \mathbb{R}$ and $t_0 \in I$. We could picture this by using F to define a slope field, as shown in Fig. 5.6. The graph of a solution $(t, u(t))$ must be parallel to the slope field at each point of the curve.

To solve (5.33) we will use a corresponding integral equation. Under the assumption that F is continuous, the two parts of the fundamental theorem of calculus immediately yield the following:

Lemma 5.40 *If $u : I \to \mathbb{R}$ is a continuous function satisfying*

$$u(t) = u_0 + \int_{t_0}^{t} F(s, u(s)) \, ds, \tag{5.34}$$

then u is C^1 and satisfies (5.33). Conversely, if u is a C^1 solution of (5.33), then u satisfies (5.34).

The strategy to solve (5.34) is to think of the right-hand side as the image of a map $u \mapsto Tu$ given by

$$Tu(t) := u_0 + \int_{t_0}^{t} F(u(s), s) \, ds. \tag{5.35}$$

A solution of (5.34) is a fixed point for which $u = Tu$. Our goal is use the fixed point theorem introduced in Exercise 4.19 to show that a solution exists and is unique.

Some additional assumption on the function F beyond continuity is required for this result to be true. For example, if $I = \mathbb{R}$ and

$$F(u, t) = 2|u|^{1/2},$$

5.5 Picard Iteration

Fig. 5.7 A continuous slope field $F(t, u)$ with multiple solutions passing through the same point

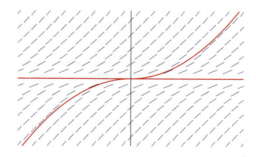

Then for $u_0 = 0$ and $t_0 = 0$ we have four possible C^1 solutions, with $u(t)$ given by either 0 or t^2 for t positive and 0 or $-t^2$ for t negative. This failure of uniqueness is illustrated in Fig. 5.7.

Another issue to keep in mind is that solutions may not exist for all times. For example, let $F(t, u) = -u^2$. For $u_0 = 1/c > 0$ and $t_0 = 0$, the equation has a unique solution $u(t) = (c - t)^{-1}$, which blows up as $t \to c$.

The metric space on which we will frame this contraction argument is $C(J; K)$, the space of continuous maps $J \to K$ where J and K are compact intervals in $\mathbb{R}$. Since $C(J; \mathbb{C})$ is complete with respect to the max-norm topology by Exercise 4.23 and $C(J; K)$ is a closed subspace, $C(J; K)$ is complete.

Theorem 5.41 (Picard) *Let $F : I \times \mathbb{R} \to \mathbb{R}$ be continuous, where I is a compact interval. For $u_0 \in \mathbb{R}$ and $r > 0$, suppose there exists $c > 0$ so that F satisfies the Lipschitz condition*

$$|F(t, u) - F(t, v)| \leq c|u - v|$$

for $u, v \in [u_0 - r, u_0 + r]$ and $t \in I$. Then for $t_0 \in I$ there exists a compact interval $J \subset I$ containing t_0 such that the equation (5.33) admits a unique solution for $t \in J$.

Proof Given u_0 and t_0, let T be the map (5.35) and set $K = [u_0 - r, u_0 + r]$. For some $a > 0$ to be chosen later we set

$$J = [t_0 - a, t_0 + a] \cap I.$$

For $u \in C(J; K)$, the integral estimate (5.30) gives

$$|Tu(t) - u_0| \leq M|t - t_0|, \tag{5.36}$$

where

$$M := \sup_{J \times K} |F(\cdot, \cdot)|,$$

which is finite because F is continuous and $J \times K$ is compact. Thus, (5.36) implies that $Tu(t) \in K$ for $|t - t_0| \leq r/M$. Since Tu is continuous, this shows that T maps $C(J; K)$ to itself provided $a < r/M$.

For $u, v \in C(J; K)$ the Lipschitz assumption on F implies that

$$|Tu(t) - Tv(t)| \leq \int_{t_0}^{t} |F(u(s), s) - F(v(s), s)| \, ds$$
$$\leq c \int_{t_0}^{t} |u(s) - v(s)| \, ds.$$

With t restricted to J this gives

$$\|Tu - Tv\|_{\max} \leq ca \|u - v\|_{\max}.$$

Thus, if we assume that $a < 1/c$ in addition to $a < r/M$, then T is a contraction on $C(J; K)$. By Exercise 4.19, there exists a unique function $u \in C(J; K)$ such that $Tu = u$, and hence a unique solution of (5.33) for $t \in J$. □

Solutions to Exercises

Chapter 1

1.5 Suppose $x_n \to y \in \mathbb{R}$. Give $\varepsilon > 0$, pick N so that $n \geq N$ implies $|x_n - y| < \varepsilon$. We can then set
$$M = \max\left\{|x_1|, \ldots, |x_N|, |y| + \varepsilon\right\}$$
to obtain $|x_n| \leq M$ for all n.

1.6 The sum formula follows immediately from the triangle inequality,
$$|(x_n + y_n) - (a + b)| \leq |x_n - a| + |y_n - b|.$$

For the product, we need to use the fact that one of the sequences is bounded, say $|y_n| \leq M$ for all n. Then
$$|x_n y_n - ab| \leq |(x_n - a)y_n| + |a(y_n - b)|$$
$$\leq M|x_n - a| + |a||y_n - b|.$$

1.7 Assume first that $\sup A$ is finite. For each $k \in \mathbb{N}$, $(\sup A - 1/k)$ is not an upper bound for A and so there exists $x_k \in A$ such that
$$x_k > \sup A - \frac{1}{k},$$
Since $x_k \leq \sup A$ also, this implies that $x_k \to \sup A$.

If $\sup A = \infty$, then for each k we can choose $x_k \in A$ so that $x_k \geq k$.

1.8 The infinite cases are essentially trivial, so let us assume the limits are real. Let $x = \lim x_n$ and $y = \lim y_n$. Given $\varepsilon > 0$, for n sufficiently large we have
$$x_n \geq x - \varepsilon, \qquad y_n \leq y + \varepsilon.$$

This shows that
$$x \leq y + 2\varepsilon.$$
Since ε was arbitrary, we conclude that $x \leq y$.

1.16 Let $\alpha = \limsup x_n$ and set $y_n = \sup_{k \geq n} x_k$ so that $y_n \to \alpha$. If $c > \alpha$ then the fact $y_n \to \alpha$ implies $y_n \leq c$ for some m. And therefore that $x_k \leq c$ for all $k \geq n$, which proves (i). If $c < \alpha$, then $y_n > c$ for all n since (y_n) is decreasing. Hence, for every n there exists $k \geq n$ such that $x_k > c$, which proves (ii).

Now suppose α satisfies (i) and (ii). If $\alpha = -\infty$ then (i) implies $x_n \to -\infty$. And if $\alpha = \infty$, then (ii) implies $\limsup x_n = \infty$. This leaves the case where $\alpha \in \mathbb{R}$. Let y_n be defined as above. Given $\varepsilon > 0$, (ii) implies that $y_n > \alpha - \varepsilon$ for all n. And by (i) we can choose N so that $x_k \leq \alpha + \varepsilon$ for $k \geq N$. For $n \geq N$ we have $|y_n - \alpha| \leq \varepsilon$. This shows that $y_n \to \alpha$ and therefore $\alpha = \limsup x_n$.

1.18 The proofs are essentially the same, so we consider only the lim sup case. Let $\alpha = \limsup x_n$ and $\beta = \limsup y_n$. If $\beta < \alpha$, then by Exercise 1.16 part (ii) there are infinitely $x_n > \beta$. Part (i) then implies that $\alpha \geq \beta$.

1.19 Let $a = \limsup x_n$ and $b = \limsup y_n$, both real since the sequences are bounded. For $\varepsilon > 0$, there are only finitely many n for which $x_n > a + \varepsilon$ and $y_n > b + \varepsilon$, by part (i) of Exercise 1.16. Therefore $x_n + y_n > a + b + 2\varepsilon$ for only finitely many n. Again by part (i) this implies that
$$\limsup(x_n + y_n) \leq a + b + 2\varepsilon.$$
Since ε was arbitrary, this completes the proof.

1.20 Let $b = \limsup y_n$. If either a or b equals 0, then it is easy to see that $x_n y_n \to 0$, because the sequences are positive and bounded. Therefore we can assume $a, b > 0$.

Suppose $c > ab$. Since $a < c/b$, we can choose $a_1 \in (a, c/b)$ and then $b_1 \in (b, c/a_1)$. This gives $a_1 > a$ and $b_1 > b$ such that $a_1 b_1 < c$. Since $x_n \leq a_1$ for all but finitely many n and $y_n \leq b_1$ for all but finitely many n, it follows that $x_n y_n > c$ for only finitely many n.

Now suppose $c < ab$. As above we can choose $0 < a_2 < a$ and $0 < b_2 < b$ so that $a_2 b_2 > c$. Since $x_n > a_2$ for all but finitely many n, and $y_n > b_2$ for infinitely many n, we conclude that $x_n y_n > c$ for infinitely many n.

It now follows from Exercise 1.16 that $\limsup x_n y_n = ab$.

1.21 If $\limsup x_k = -\infty$ then $x_n \to -\infty$. If $\limsup x_n = \infty$, then for each n we have $x_k \geq n$ for infinitely many k by Exercise 1.16. There for we can choose a subsequence with $x_{k_n} \geq n$, which gives $x_{k_n} \to \infty$.

Now suppose $\alpha = \limsup x_k$ is real. For $\varepsilon > 0$, Exercise 1.16 implies that there are infinitely many k for which $x_k \in (\alpha - \varepsilon, \alpha + \varepsilon]$. Hence, we can choose a subsequence such that
$$|x_{k_n} - \alpha| \leq \frac{1}{n}$$
for each n.

A Solutions to Exercises

1.26 Suppose (x_k) is a Cauchy sequence in $\mathbb{R}$. Given ε, all but finitely many x_k lie within an interval of width ε. Therefore (x_k) is bounded and so the Bolzano-Weierstrass property yields a convergent subsequence, say $x_{k_n} \to y$. Given $\varepsilon > 0$, choose N so that $|x_i - x_j| \leq \varepsilon$ for $i, j \geq N$. Then taking $n \to \infty$ in the inequality

$$|x_j - y| \leq |x_j - x_{k_n}| + |x_{k_n} - y|$$

gives

$$|x_j - y| \leq \varepsilon$$

for $j \geq N$. This proves $x_j \to y$ as $j \to \infty$.

Chapter 2

2.1 If $z_k \to z$, then the convergence of real and imaginary parts follows immediately from the inequalities

$$|x_k - x| \leq |z_k - z|, \qquad |y_k - y| \leq |z_k - z|.$$

For the other direction, assume that $x_k \to x$ and $y_k \to y$. Given $\varepsilon > 0$, we know that

$$|x_k - x| < \varepsilon \quad \text{and} \quad |y_k - y| < \varepsilon$$

for all but finitely many k. This implies that

$$|z_k - z| \leq \sqrt{2}\varepsilon$$

for all but finitely many k. Therefore $z_k \to z$.

2.3 et (z_k) be a Cauchy sequence in $\mathbb{C}$, with $z_k = x_k + iy_k$. Since

$$|x_m - x_n| \leq |z_n - z_m|,$$

the sequence (x_k) is Cauchy in $\mathbb{R}$ and therefore there $x_k \to x$ for some $x \in \mathbb{R}$. Similarly, $y_k \to y$ for some $y \in \mathbb{R}$. Setting $z = x + iy$ then gives $z_k \to z$ in $\mathbb{C}$, by Exercise 2.1.

2.4 Let $s_n := \sum_{k=1}^{n} x_k$, and suppose $s_n \to w$. Then $z_k = s_k - s_{k-1}$, and by Lemma 2.2

$$\lim_{k \to \infty} (s_k - s_{k-1}) = w - w = 0.$$

2.10 Suppose that $\sum z_k$ is absolutely convergent. By the triangle inequality,

$$|z_n + \cdots + z_m| \leq \sum_{k=n}^{m} |z_k|, \tag{A.1}$$

for $m > n$. Since $\sum |z_k| < \infty$ the right-hand side of (A.1) can be made arbitrarily small by taking $m \geq n \geq N$ for N sufficiently large. Hence $\sum z_k$ converges by Theorem 2.8.

2.13 Let α_n, β_n denote the partial sums of the series $\sum a_k$ and $\sum b_k$, and γ_m the partial sum for $\sum c_n$. Comparing terms gives

$$\gamma_{2n} = \alpha_n \beta_n + \sum_{j=n+1}^{2n} \sum_{i=0}^{2n-j} a_i b_j + \sum_{i=n+1}^{2n} \sum_{j=0}^{2n-j} a_i b_j.$$

Thus, using the absolute convergence, we can estimate

$$|\gamma_{2n} - \alpha_n \beta_n| \leq \sum_{i=0}^{\infty} |a_i| \sum_{j=n+1}^{\infty} |b_j| + \sum_{i=n+1}^{\infty} |a_i| \sum_{j=0}^{\infty} |b_j|$$

and note that the right-hand side approaches 0 as $n \to \infty$.

2.14 Let $B = \sum_{k=1}^{\infty} b_k$ and define

$$s_n := |z_1| + \cdots + |z_n|.$$

Since all partial sums for $\sum b_k$ are bounded by B, the hypothesis implies that

$$s_n \leq |z_1| + \cdots + |z_n| + B$$

for all n. Theorem 2.9 implies that $\sum |z_k|$ is convergent.

2.17 Suppose

$$\alpha = \limsup_{k \to \infty} \left| \frac{z_{k+1}}{z_k} \right| < 1.$$

Then for $\alpha < r < 1$ there exists N such that

$$k \geq N \implies \left| \frac{z_{k+1}}{z_k} \right| \leq r.$$

This implies that

$$z_k \leq C_N r^k$$

for all $k \geq N$. Convergence then follows from Exercise 2.14.

Chapter 3

3.5 For $\alpha \in \ell^1(\mathbb{N})$ we have $|\alpha_j| \leq \|\alpha\|_\infty$ for all j. We can thus estimate

$$\|\alpha\|_p = \left(\sum_j |\alpha_j| \cdot |\alpha_j|^{p-1} \right)^{\frac{1}{p}}$$

$$\leq \|\alpha\|_\infty^{1-\frac{1}{p}} \|\alpha\|_1^{\frac{1}{p}}.$$

A Solutions to Exercises

Taking $p \to \infty$ thus gives
$$\limsup_{p \to \infty} \|\alpha\|_p \leq \|\alpha\|_\infty.$$

To estimate from below, let $c < \|\alpha\|_\infty$. Then we know that $|\alpha_j| \geq c$ for at least one j. This implies that
$$\|\alpha\|_p \geq c$$
for $p \geq 1$, and so
$$\liminf_{p \to \infty} \|\alpha\|_p \geq c.$$

Since c was arbitrary, we conclude that
$$\liminf_{p \to \infty} \|\alpha\|_p \geq \|\alpha\|_\infty.$$

3.8 (a) False. The endpoints of a closed interval in $\mathbb{R}$ are neither interior nor isolated.
   (b) True. Suppose $x \in E$ is not interior. Then for every $\varepsilon > 0$, $N_\varepsilon(x)$ is not contained in E. Since $x \in E$, this implies that x is a boundary point.
   (c) True. If x is not interior to either E of E^c, then for all $\varepsilon > 0$, $N_\varepsilon(x)$ is not contained in either E or E^c. This implies that $x \in \partial E$.
   (d) True. If $x \in E$ is not isolated, then for all $\varepsilon > 0$ the neighborhood $N_\varepsilon(x)$ contains some point of E other than x. Thus x is a limit point.
   (e) True. If x is a limit point of E then $N_\varepsilon(x)$ contains other points of E for all $\varepsilon > 0$. If x is not interior, then all of these $N_\varepsilon(x)$ intersect E^c, implying that x is a boundary point.

3.9 It is clear from the definitions that an accumulation point is a limit point. Suppose that x is a limit point of E. Given $\varepsilon > 0$ there exists some $x_1 \in N_\varepsilon(x)$ not equal to x. We can then create a sequence inductively by setting $\varepsilon_n = d(x, x_n)$ and choosing x_{n+1} as a point in $N_{\varepsilon_n}(x) \setminus \{x\}$. This gives an infinite sequence of distinct points in $N_\varepsilon(x)$.

3.10 For $x_0 \in X$ and $r > 0$ consider $x \in N_r(x_0)$. For $\varepsilon := r - d(x, y)$, the triangle inequality implies that
$$N_\varepsilon(y) \subset N_r(x_0).$$
This proves that $N_r(x_0)$ is open.

3.11 Suppose $A \subset X$ is open. If $x \in A$ then $N_\varepsilon(x) \subset A$ for some $\varepsilon > 0$, implying that x is not a limit point of A^c. This shows that all limit points of A^c are contained in A^c, hence A^c is closed. Conversely, if A is not open then there is at least one point $x \in A$ which is not interior. This means for each $\varepsilon > 0$ there is at least one point of A^c in $N_\varepsilon(x)$. Hence x is a limit point of A^c, which shows that A^c is not closed.

3.12 (a) Let $A = \cup_\alpha U_\alpha$ where each U_α is open. If $x \in A$ then $x \in U_\alpha$ for some α and there exists a neighborhood $N_\varepsilon(x) \subset U_\alpha$. This gives $N_\varepsilon(x) \subset A$.
   (b) Let $A = \cap_{j=1}^n U_j$ where U_j is open. For $x \in A$, there exist neighborhoods $N_{\varepsilon_j}(x) \subset U_j$ for each j. Taking $\varepsilon = \min(\varepsilon_1, \ldots, \varepsilon_n)$ gives $N_\varepsilon(x) \subset A$.

(c) This follows immediately from (a) and (b).
(d) This follows immediately from parts (a) and (c).

3.14 It suffices to prove that if a limit point of E is either contained in E or a boundary point. Suppose x is a limit point of E and $x \notin E$. Then for every $\varepsilon > 0$, $N_\varepsilon(x)$ contains at least one point of E. Since $x \in E^c$, this proves that x is a boundary point.

3.15 (a) True. Since $\overline{A} \cup \overline{B}$ is closed and contains $A \cup B$,
$$\overline{A \cup B} \subset \overline{A} \cup \overline{B}.$$
On the other hand, a limit point of either A or B is clearly also a limit point of $A \cup B$, so
$$\overline{A} \cup \overline{B} \subset \overline{A \cup B}.$$
(b) False. Let $A = (0, 1)$ and $B = (-1, 0)$ in $\mathbb{R}$, so $A \cap B = \emptyset$ but $\overline{A} \cap \overline{B} = \{0\}$.
(c) False. Let $A = \mathbb{R} \setminus \{0\}$ and $B = \{0\}$ in $\mathbb{R}$.
(d) False. Let $A = \mathbb{R}$ and $B = \mathbb{N}$ in $\mathbb{R}$.

3.16 Suppose $E \subset X$ is dense, and let U be a non-empty open set. For $x \in U$, there exists a neighborhood $N_r(x) \subset U$. Since x is a limit point of E, this neighborhood must contain some point of E.

Now suppose that E intersects every non-empty open set of X. Given $x \in X$ and $\varepsilon > 0$, the set $N_\varepsilon(x) \setminus \{x\}$ is open and therefore intersects E. This shows that for all $\varepsilon > 0$ $N_\varepsilon(x)$ contains a point of E other than x, proving that x is a limit point.

3.18 Suppose E is connected and $E = A \cap B$ where $\overline{A} \cap B$ and $A \cap \overline{B}$ are both empty. If $x \in E$ is a limit point of A, then $x \in \overline{A}$ and thus $x \notin B$. Therefore $x \in A$ since $E = A \cap B$. This shows A is closed relative to E. The same argument shows that B is closed relative to E. Therefore A and B are also both open relative to E by Exercise 3.11. Hence either A or B is empty since E is connected.

Now suppose that $E = A \cap B$ where A and B are non-empty separated sets. The argument used above shows that A and B are both open and closed relative to E, so E is not connected.

3.19 Suppose A and B are disjoint open sets. For $x \in A$, the existence of a neighborhood $N_\varepsilon(x) \subset A$ implies that x is not a limit point of B. Hence $A \cap \overline{B} = \emptyset$. By the same reasoning, $\overline{A} \cap B = \emptyset$.

3.21 Suppose x_0 is a limit point of E. For each k, there exists a point in $E \setminus \{x_0\}$ with $|x_k - x_0| \leq 1/k$. This gives $x_k \to x_0$.

Conversely, if $x_k \in E \setminus \{x_0\}$ and $x_k \to x_0$, then for each $\varepsilon > 0$ the interval $(x_0 - \varepsilon, x_0 + \varepsilon)$ contains infinitely many x_k. Hence x_0 is a limit point.

3.23 Given $\varepsilon > 0$, choose N so that $d(x_n, x_m) \leq \varepsilon$ for $n, m \geq N$. By the triangle inequality,
$$d(x_n, x) \leq d(x_n, x_{n_k}) + d(x_{n_k}, x).$$
For $n \geq N$ we can take $k \to \infty$ to obtain

$$d(x_n, x) \leq \varepsilon.$$

This proves that $x_n \to x$.

3.24 Suppose Y contains a Cauchy sequence (y_k). Since Y is a metric subspace of X, (y_k) is also Cauchy in X. Hence y_k converges at some point $x \in X$. Since Y is closed relative to X, we have $x \in Y$. Therefore Y is complete.

On the other hand, suppose Y is complete. If x is a limit point of Y, then there exists a sequence of $y_k \in Y$ such that $y_k \to x$ in X. This implies that (y_k) is Cauchy in X. The sequence is also Cauchy in Y under the relative topology, so (y_k) is convergent in Y also. Since x is the limit of the sequence, this shows that $x \in Y$. Hence Y is closed.

3.30 Suppose (α^n) is a Cauchy sequence in $\ell^\infty(\mathbb{N})$, with $\alpha^n = (\alpha_1^n, \alpha_2^n, \ldots)$. This implies that $(\alpha_j^n)_{n \in \mathbb{N}}$ is Cauchy in $\mathbb{C}$ for each j, since

$$|\alpha_j^n - \alpha_j^m| \leq \|\alpha^n - \alpha^m\|_\infty$$

We can thus set
$$\beta_j = \lim_{n \to \infty} \alpha_j^n$$
to define the target sequence β. We can see that $\beta \in \ell^\infty$ by

$$|\beta_j| \leq \sup_n \|\alpha^n\|_\infty,$$

which is finite by the Cauchy assumption. It remains to show that $\alpha^n \to \beta$ in ℓ^∞.

Given $\varepsilon > 0$, choose N so that $n, m \geq N$ implies $\|\alpha^n - \alpha^m\|_\infty \leq \varepsilon$, which implies

$$|\alpha_j^n - \alpha_j^m| \leq \varepsilon$$

for all j. Taking $m \to \infty$ then yields

$$|\alpha_j^n - \beta_j| \leq \varepsilon$$

for $n \geq N$ and all j. Hence $\|\alpha^n - \beta\|_\infty \leq \varepsilon$ for $n \geq N$.

3.32 For $a_k := d(x_k, y_k)$ our goal is to show that (a_k) is a Cauchy sequence in $\mathbb{R}$. Given $\varepsilon > 0$, choose N_1 so that $d(x_k, x_l) \leq \varepsilon$ for $k, l \geq N_1$. Choose N_2 so that $d(y_k, y_l) \leq \varepsilon$ for $k, l \geq N_2$. Then, for $n \geq N := \max(N_1, N_2)$, the triangle inequality gives

$$|a_n - d(x_N, y_N)| \leq d(x_n, x_N) + d(y_n, y_N)$$
$$\leq 2\varepsilon.$$

For $n, m \geq N$ this implies
$$|a_n - a_m| \leq 4\varepsilon.$$

This proves that the sequence (a_k) is Cauchy, and hence a limit exists.

Let (x'_n) and (y'_n) be Cauchy sequences equivalent to (x_n) and (y_n), respectively. By the triangle inequality,

$$|d(x'_n, y'_n) - d(x_n, y_n)| \leq d(x'_n, x_n) + d(y'_n, y_n).$$

This implies that
$$\lim_{n \to \infty} |d(x'_n, y'_n) - d(x_n, y_n)| = 0.$$

3.34 Let $\alpha \in X^*$ be represented by the Cauchy sequence (x_k) in X. Given $\varepsilon > 0$, our goal is to produce an element $y \in X$ such that $d^*(\alpha, [(y)]) \leq \varepsilon$. Choose N so that $d(x_k, x_l) \leq \varepsilon$ for all $k, l \geq N$ and set $y = x_N$. Then,

$$d(x_k, y) \leq \varepsilon$$

for all $k \geq N$. Hence, taking the limit as $k \to \infty$ gives

$$d^*(\alpha, [(y)]) \leq \varepsilon.$$

3.38 Suppose first that K is compact relative to Y. Let $\{U_\alpha\}$ be an open over of K in Y. By Theorem 3.17, the sets $V_\alpha := U_\alpha \cap Y$ are open in Y. Since $K \subset Y$, $\{V_\alpha\}$ is an open cover of K in Y. By compactness in Y, there exists a finite subcover

$$K \subset \bigcup_{j=1}^{m} V_j.$$

Then $\cup_{j=1}^{m} U_j$ gives a finite subcover of K in X.

Now suppose that K is compact relative to X, and let $\{V_\alpha\}$ be an open cover of X in Y. By Theorem 3.17 each V_α can be written as $U_\alpha \cup Y$ for some open set U_α in X. Compactness gives a finite subcover

$$K \subset \bigcup_{j=1}^{m} U_j,$$

and since $K \subset Y$, this implies that $K \subset \cup_{j=1}^{m} V_j$.

3.42 Suppose, for the sake of contradiction, that $\cap K_j = \emptyset$. For $U_j := K_j^c$, then this implies $X = \cap U_j$. In particular, $\{U_j\}$ gives an open cover for K_1. By compactness,

$$K_1 \subset U_1 \cup \cdots \cup U_m$$

for some finite m. By the nesting, the union on the right equals U_m, implying that K_1 and K_m are disjoint. This contradicts the hypotheses, since $K_m \subset K_1$ and K_m is not empty.

3.47 Suppose that K is sequentially compact, and let x be a limit point of K. This means that there exists a sequence (x_k) in K converging to x. Since every subsequence converges to x also, sequential compactness implies that $x \in K$.

A Solutions to Exercises

Assume, for the sake of contradiction, that K is unbounded. Then for some fixed $x_0 \in X$, we can choose a sequence (x_k) in K such that $d(x_0, x_k) \to \infty$ as $k \to \infty$. Since any subsequence also has this property, there can be no convergent subsequence.

3.48 Let F be a closed and bounded subset of $\mathbb{R}^n$ containing a sequence (x_k). Since (x_k) is bounded, it admits a convergent subsequence by Bolzano-Weierstrass (Theorem 3.26). The limit point is contained in F because F is closed.

3.49 Suppose K is compact and F is closed, with $d(K, F) = 0$. Then there exist sequences (x_n) in K and (y_n) in F such that $d(x_n, y_n) \to 0$. By Theorem 3.44, there is a subsequence (x_{n_k}) converging to some $x \in K$. This implies $y_{n_k} \to x$ also, which means $x \in F$ since F is closed. Thus K and F are not disjoint.

3.53 Suppose that F is closed. Note that the definition of a boundary point is symmetric between a set and its complement, so that $\partial F = \partial(F^c)$. Since F is closed and F^c is open, this means that $\partial F \subset F$ and
$$\overline{F^c} = F^c \cup \partial F.$$
Hence F^c is dense if and only if $\partial F = F$. The latter condition means precisely that F has empty interior.

Chapter 4

4.4 Consider two points $x_1, x_2 \in X$. Given $\varepsilon > 0$ we can find $y \in A$ such that
$$d(y, x_1) \leq d(A, x_1) + \varepsilon,$$
since $d(A, \cdot)$ is defined as an infimum. By the triangle inequality,
$$d(y, x_2) \leq d(y, x_1) + d(x_1, x_2)$$
$$\leq d(A, x_1) + d(x_1, x_2) + \varepsilon.$$
This implies that
$$d(A, x_2) \leq d(A, x_1) + d(x_1, x_2) + \varepsilon,$$
and thus, since ε was arbitrary
$$d(A, x_2) \leq d(A, x_1) + d(x_1, x_2).$$
Applying the same argument with the points switched gives
$$|d(A, x_2) - d(A, x_1)| \leq d(x_1, x_2).$$

4.5 If x is a limit point of E, then there exists a sequence (x_k) in E converging to x. By Lemma 4.7, $f(x_k)$ converges to $f(x)$. Since $f(x_k) \in f(E)$, this proves that $f(x) \in \overline{f(E)}$.

4.6 Suppose first that f is continuous and that $x_n \to a$. Given $\varepsilon > 0$ continuity implies that there exists $\delta > 0$ so that $d(f(x), f(a))$ for $d(x, a) \leq \delta$. The convergence of the sequence gives N so that $d(x_n, a) \leq \delta$ for all $n \geq N$. and this completes the proof for this direction.

To prove the converse statement, suppose that f is not continuous at a. This means that that a is a limit point of E and there exists an $\varepsilon > 0$ for which the continuity condition (4.1) is not satisfied for any choice of δ. In particular, we can set $\delta = 1/n$ for $n \in \mathbb{N}$ to form a sequence $x_n \in E \setminus \{a\}$ such that $d(x_n, a) \leq 1/n$ and $d(f(x_n), f(a)) > \varepsilon$. This gives $x_n \to a$ with $f(x_n) \not\to f(a)$.

4.11 Suppose $f : X \to Y$ is continuous and $K \in X$ is compact. Let $\{U_\alpha\}$ be an open cover for $f(K)$. Then $\{f^{-1}(U_\alpha)\}$ is an open cover for K. There exists a finite subcover

$$K \subset \bigcup_{j=1}^{m} f^{-1}(U_j),$$

which implies that $f(K)$ has the finite subcover $\cup_{j=1}^{m} U_j$.

For the sequential proof, assume that (y_n) is a convergence sequence in $f(K)$. For each n choose $x_n \in K$ so that $f(x_n) = y_n$. Then by compactness (x_n) has a subsequence x_{n_k} converging to some $x \in K$ as $k \to \infty$. By continuity, $y_{n_k} \to f(x)$ as $k \to \infty$. This shows that $f(K)$ is sequentially compact.

4.14 Let $f : E \to Y$ be a continuous map, with E connected. Suppose $A \subset f(E)$ is both open and closed relative to $f(E)$. Then, by Theorem 4.8 and its corollary, $f^{-1}(A)$ is open and closed relative to E. Therefore, either $f^{-1}(A) = E$ or $f^{-1}(A) = \emptyset$. This implies that either $A = f(E)$ or $A = \emptyset$. Thus $f(E)$ is connected.

4.17 Suppose that f is continuous but not uniformly continuous on K. Then there exists an $\varepsilon > 0$ such that at least one exception to (4.4) exists for every $\delta > 0$. In particular, for each $n \in \mathbb{N}$ we can find points x_n, y_n in X with

$$d(x_n, y_n) < \frac{1}{n} \quad \text{and} \quad d(f(x_n), f(y_n)) \geq \varepsilon. \tag{A.2}$$

Since K is compact, there exists a subsequence (x_{n_k}) converging to some $a \in X$. By (A.2), $y_{n_k} \to a$ also. The continuity of f then implies that

$$\lim_{k \to \infty} f(x_{n_k}) = \lim_{k \to \infty} f(y_{n_k}) = f(a),$$

which contradicts (A.2). We conclude that f is uniformly continuous.

4.18 For each n, choose δ_n so that

$$|x - y| \leq \delta_n \quad \Rightarrow \quad |f(x) - f(y)| \leq \frac{1}{n}.$$

This shows that for any sequence with $x_n \in (0, \delta_n)$, the sequence $f(x_n)$ is Cauchy and therefore has a limit. Moreover, given another such sequence y_n,

$$|f(x_n) - f(y_n)| \le \frac{2}{n},$$

so the two image sequences have the same limit. It follows that $\lim_{x \to 0} f(x)$ exists. The same reasoning applies to $x \to 1$.

4.22 Given $\varepsilon > 0$, by uniform convergence there exists n sufficiently large so that

$$\sup_{x \in E} d(f_n(x), f(x)) < \varepsilon.$$

Fixing this value of n, we use the continuity of f_n to choose $\delta > 0$ so that

$$d(x, q) < \delta \implies d(f_n(x), f_n(q)) < \varepsilon.$$

For $x \in E$ with $d(x, q) < \delta$, the triangle inequality now gives

$$d(f(x), f(q)) \le d(f(x), f_n(x)) + d(f_n(x), f_n(q)) + d(f_n(q), f(q))$$
$$< 3\varepsilon.$$

Since ε was arbitrary, this shows f is continuous at q.

4.23 Suppose (f_k) is a Cauchy sequence in $C(X, \mathbb{C})$. This implies that $(f_k(x))$ is Cauchy in $\mathbb{C}$ for each x, so we can define a function f by the pointwise limit

$$f(x) := \lim_{k \to \infty} f_k(x).$$

The Cauchy condition means that given $\varepsilon > 0$, there exists N so that

$$|f_k(x) - f_m(x)| \le \varepsilon$$

for all $k, m \ge N$ and $x \in X$. With x fixed we can take the limit $m \to \infty$ here, to obtain

$$|f_k(x) - f(x)| \le \varepsilon$$

for all $k \ge N$. This implies that $f_k \to f$ uniformly, which also shows that $f \in C(X, \mathbb{C})$.

Chapter 5

5.3 For $|x| \ge 1$ we have $|x|^k \le |x|^n$ for $k \le n$, and so

$$|p(x)| \le |a_n x^n| + \cdots + |a_1 x|$$
$$\le \left(|a_n| + \cdots + |a_1|\right)|x|^n.$$

Similarly, for $|x| \le 1$

$$|p(x)| \le \left(|a_n| + \cdots + |a_1|\right)|x|.$$

5.4 For $|x| < R$ we can write
$$f(x) - \sum_{n=0}^{m-1} a_n x^n = \sum_{n=m}^{\infty} a_n x^n = x^m g(x),$$
where
$$g(x) = \sum_{k=0}^{\infty} a_{k+m} x^k.$$
Since the power series for $x^m g(x)$ agrees with that of $f(x)$ up to finitely many terms, g has the same radius of convergence and therefore represents a continuous function on $(-R, R)$. For $r < R$ we can set
$$C = \sup_{|x| \leq r} |g(x)|,$$
which gives
$$\left| f(x) - \sum_{n=0}^{m-1} a_n x^n \right| \leq C |x|^m.$$

5.5 Setting $r(x) = f(x) - \phi(x)$ gives
$$\frac{f(x)}{\phi(x)} - 1 = \frac{r(x)}{\phi(x)}.$$
The limit on the left exists and equals zero if and only if the limit on the right exists and equals zero.

5.8 Suppose first that f is continuous. Then f is bounded by the extreme value theorem, and hence $\Gamma(f)$ is a bounded subset of $\mathbb{R}^2$. Let (x, y) be a limit point of $\Gamma(f)$, so that there exists a sequence
$$(x_n, f(x_n)) \to (x, y)$$
in $\mathbb{R}^2$. In particular, this means that $x_n \to x$ and $f(x_n) \to y$. Since f is continuous, $f(x) = \lim f(x_n)$ also, and so $f(x) = y$. This shows that $(x, y) \in \Gamma(f)$. Therefore $\Gamma(f)$ is closed.

Now suppose that f is discontinuous at x. Then for some $\varepsilon > 0$ we can find for each n a point x_n with $|x - x_n| < 1/n$ and $|f(x_n) - f(x)| \geq \varepsilon$. Since any subsequence of (x_n) converges to x, the only possible limit point for a subsequence of $(x_n, f(x_n))$ in $\Gamma(f)$ is $(x, f(x))$. However, for all n we have
$$\|(x_n, f(x_n)) - (x, f(x))\| \geq \varepsilon,$$
and thus there is no convergent subsequence. This shows that $\Gamma(f)$ is not compact.

5.9 It suffices to assume that f is increasing. Suppose x is an interior point of I, and choose $\varepsilon > 0$ so that $N_\varepsilon(x) \subset I$. Then we can define
$$f_+(x) := \sup_{(x-\varepsilon, x)} f, \quad f_-(x) := \inf_{(x, x+\varepsilon)} f$$

By the monotone convergence theorem it is easy to check that the left and right limits exist and are given by
$$\lim_{t \to x^\pm} f(t) = f_\pm(x).$$
Therefore the only possible discontinuities of f are jumps.

For each jump discontinuity, we can choose a rational number that lies strictly between the left and right limits. Because f is monotonic, this assigns a distinct rational number to each jump. Therefore the number of jumps is countable.

5.12 Suppose $f : I \to \mathbb{R}$ is differentiable at $x \in I$. For small $t \neq 0$ we can write
$$f(x+t) - f(x) = \left(\frac{f(x+t) - f(x)}{t}\right) t.$$
Since both terms on the right have limits as $t \to 0$, we can take the limits separately by Lemma 5.1. This gives
$$\lim_{\varepsilon \to 0}[f(x+\varepsilon) - f(x)] = 0,$$
proving that f is continuous at x.

5.13 The additive property follows immediately from Lemma 5.1. For the product rule, we start from a simple algebraic computation
$$\frac{fg(x+\varepsilon) - fg(x)}{\varepsilon} = \left(\frac{f(x+\varepsilon) - f(x)}{\varepsilon}\right) g(x+\varepsilon) + f(x)\left(\frac{g(x+\varepsilon) - g(x)}{\varepsilon}\right).$$
Since the limits as $\varepsilon \to 0$ exist separately for all terms on the right, we can take $\varepsilon \to 0$ to conclude that
$$(fg)'(x) = f'(x)g(x) + f(x)g'(x).$$

5.15 Suppose that f is convex on I. For $x, y \in I$ we can rewrite the convexity inequality (5.12) as
$$t[f(y) - f(x)] \geq f(x + t(y-x)) - f(x).$$
Replacing t by $h = t(y-x)$ gives
$$f(y) - f(x) \geq \frac{f(x+h) - f(x)}{h}(y-x).$$
By differentiability we can take $h \to 0$ to obtain
$$f(y) - f(x) \geq f'(x)(y-x).$$
This proves that curve lies above the tangent line at x.

Now suppose that the tangent inequality (5.13) holds at each point. Given $x, y \in I$ and $t \in [0, 1]$ we let x_t be the interpolated point
$$x_t = (1-t)x + ty.$$

The tangent inequalities at x_t then give

$$f(x) \geq f(x_t) + f'(x_t)t(x - y),$$
$$f(y) \geq f(x_t) + f'(x_t)(1 - t)(y - x).$$

If we multiply the first line by $1 - t$ and the second by t, then the $f'(x_t)$ terms cancel when we add these together, leaving

$$(1 - t)f(x) + tf(y) \geq f(x_t).$$

This proves that f is convex.

5.21 If x_0 is a local extremum, then the difference quotient

$$\frac{f(x) - f(x_0)}{x - x_0}$$

changes sign at x_0, by either (5.19) or (5.19). Since the limit $x \to x_0$ exists by assumption, this implies

$$\lim_{x \to x_0} \frac{f(x) - f(x_0)}{x - x_0} = 0.$$

5.35 Given $\varepsilon > 0$, choose f_n so that

$$\|f - f_n\|_{\max} < \varepsilon.$$

This implies that for any partition P,

$$|S^{\pm}(f, P) - S^{\pm}(f_n, P)| < (b - a)\varepsilon.$$

Since f_n is integrable, there exists a partition P_n such that

$$S^{+}(f_n, P_n) - S^{-}(f_n, P_n) < (b - a)\varepsilon.$$

Replacing f_n by f gives

$$S^{+}(f, P_n) - S^{-}(f, P_n) < 3(b - a)\varepsilon,$$

which shows that

$$\inf_P S^{+}(f, P) - \sup S^{-}(f, P) \leq 3(b - a)\varepsilon.$$

Since ε was arbitrary, this proves that f is integrable. It then follows from (5.31) that

$$\left|\int_a^b (f_n - f)\right| \leq (b - a)\|f - f_n\|_{\max},$$

so uniform convergence implies convergence of the integrals.

5.36 Clearly $\|\cdot\|$ is positive and homogeneous, and the triangle inequality follows immediately from monotonicity. The only norm property that remains to be checked is definiteness. Suppose that $f \in V$ satisfies $|f(x_0)| > 0$ at some point x_0, and set $c = |f(x_0)|/2$. By continuity the set $\{x : |f(x)| > c\}$ is open, and so there exists $\varepsilon > 0$ such that

$$|f(x)| > c \quad \text{for } |x - x_0| < \varepsilon.$$

By monotonicity this implies that

$$\int_0^1 |f| \geq 2c\varepsilon > 0.$$

Thus, if $\|f\| = 0$ then $f = 0$ at all points.

5.39 By Theorem 5.38 we can write

$$f_n(x) = f_n(a) + \int_a^x f_n'.$$

Using the hypotheses and Exercise 5.35, we can take $n \to \infty$ to deduce

$$f(x) = f(a) + \int_a^x g.$$

Since f_n' is continuous, g is also by uniform convergence. Theorem 5.37 then implies f is differentiable with $f' = g$.

Bibliography

There are many fine introductory analysis texts that give a thorough presentation of the core material, and many of these include additional topics that we have not had the space for here. Here is short selection of texts that have provided inspiration for this book.

References

Abbott, Stephen. 2015. *Understanding analysis*, 2nd ed. Springer.
Apostol, Tom M. 1974. *Mathematical analysis*, 2nd ed. Addison-Wesley.
Conway John, B. 2018. Cambridge: A first course in analysis.
Rudin, Walter. 1976. *Principles of mathematical analysis*, 3rd ed. McGraw-Hill.
Strichartz, Robert S. 1995. *The way of analysis*. Jones and Bartlett
Wade, William R. 2009. *An introduction to analysis*, 4th ed. Prentice Hall.

Index

Symbols
C^m, 83
$\mathbb{Q}$, 1
$\mathbb{R}^n$, 30
$\mathbb{R}_\infty$, 2
ℓ^p, 31, 41

A
Accumulation point, 34
Archimedean property, 3
Asymptotic, 74

B
Baire category theorem, 53
Banach fixed-point theorem, 62
Bolzano-Weierstrass theorem, 15, 40
Boundary point, 34
Bounded sequence, 6, 38
Bounded set, 46

C
Cantor set, 52
Cauchy sequence, 13, 18, 38
Closed, 5, 34
Closure, 35
Compact, 46, 48, 59
  limit-point, 49
  sequentially, 49
Completeness, 1, 13, 19, 39, 43

Connected set, 37
Continuity, 55
  Lipschitz, 61
  uniform, 60
Continuous, 55
Contraction, 62
Convergence, 38
  absolute, 21
  conditional, 21
  pointwise, 62
  uniform, 62
Convex function, 78
Countable, 9

D
Dense, 36, 45, 53, 54
Derivative, 76
Differentiable, 76
  infinitely, 83
Dini's theorem, 64
Distance function, 29

E
Euclidean norm, 30
Euler's formula, 26
Exponential series, 23
Extended real numbers, 2

F

Function limit, 57, 71
Fundamental theorem, 94

G
Geometric series, 20

H
Hölder's inequality, 32
Harmonic series, 19
　alternating, 21
Heine-Borel theorem, 48

I
Infimum, 3
Interior point, 34
Interval, 4
Isolated point, 34

J
Jump discontinuity, 76

L
lim inf, 10
lim sup, 10
Limit, 5, 38
　point, 34
　upper/lower, 10
Lower bound, 3

M
Maximum, 3
　local, 83
Max norm, 31
Meager set, 52
Metric, 29
　completion, 44
　space, 29
　subspace, 36
Minimum, 3
　local, 83
Minkowski inequality, 31
Monotone sequence, 7
Monotonic function, 76

N
Neighborhood, 33
Nested compact set property, 49
Norm, 30
　equivalence, 40
Normed vector space, 30
Nowhere dense, 52

O
Open, 4, 34
　cover, 46
　map, 59
Ordered field, 1
　complete, 3
Order notation, 73

P
Picard iteration, 95
Power series, 25

R
Radius of convergence, 25
Rational cut, 2
Ratio test, 24
Relative topology, 36
Root test, 24, 25

S
Separated sets, 37
Sequence, 5
　Cauchy, 13, 18, 38
　monotone, 7
Series, 19
　absolute convergence, 43
　exponential, 23
　geometric, 20
　harmonic, 19
　power, 25
　rearrangement, 22
Subsequence, 12
Subspace topology, 36
Supremum, 3

T
Topology, 29
Triangle inequality, 29

U
Uncountable, 9
Upper bound, 3

Y
Young's inequality, 32, 85

Printed in the United States
by Baker & Taylor Publisher Services